1983

# ELEMENTARY FUNCTIONS

# ELEMENTARY FUNCTIONS

Carl B. Allendoerfer

Cletus O. Oakley

Haverford College

Donald R. Kerr, Jr.

Indiana University

**McGraw-Hill
Book Company**

New York
St. Louis
San Francisco
Auckland
Bogota
Düsseldorf
Johannesburg
London
Madrid
Mexico
Montreal
New Delhi
Panama
Paris
São Paulo
Singapore
Sydney
Tokyo
Tornoto

This book was set in Melior by Ruttle, Shaw & Wetherill, Inc.
The editors were A. Anthony Arthur, Alice Macnow,
and Shelly Levine Langman;
the designer was J. E. O'Connor;
the production supervisor was Dennis J. Conroy.
The drawings were done by J & R Services, Inc.
R. R. Donnelley & Sons Company was printer and binder.

**ELEMENTARY FUNCTIONS**

1234567890 DODO 7832109876

**Library of Congress Cataloging in Publication Data**

Allendoerfer, Carl Barnett, date
    Elementary functions.

    1.  Functions.  I.  Oakley, Cletus Odia, date
joint author.  II.  Kerr, Donald R., joint author.
III.  Title.
QA331.3.A44     512'.1     76-10242
ISBN 0-07-001371-3

To the memory of
Carl B. Allendoerfer
whose imprint on
American Mathematics
was substantial

# CONTENTS

# PREFACE

The material in this text is an elementary approach to the subject of *elementary functions* involving the polynomial functions, the exponential and logarithmic functions, and the trigonometric functions. All of these functions play a most important role in calculus and other specialized areas of mathematics.

As a knowledge of calculus is the "open-sesame" to a study of higher mathematics, so it is that a sound grasp of elementary functions is required of a student who wishes to learn calculus. We believe that the material in this text is sufficient in kind, quantity, and quality to serve both as a precursor to further work and as a terminal course in mathematics. Although it is primarily designed as a one-semester book, most of its contents could be fitted into a one-quarter course. It should fill a need of those college freshmen (or even high school seniors) who are not quite ready to begin a serious study of the calculus. The minimum prerequisite for a course using this text would be 1 year of plane geometry and $1\frac{1}{2}$ years of the kind of school mathematics which is usually called *algebra*.

For clarity we have set off *definitions, theorems, proofs,* etc., in boldface type. Practically every kind of problem found in the problem sets was previously enunciated and solved in the various illustrations. In general the problems are graded so that the easier ones come first and the harder ones last. They are taken not only from the field of mathematics itself but also from the social, physical, and biological sciences. Altogether there are more than one thousand. Usually an even-numbered problem has its odd-numbered counterpart. Answers are provided for the odd-numbered ones.

A special feature of the book is the large number of graphs and other figures. We are indebted to Justine Baker for working through most of the problems and for her keen eye in reading proof. We are also indebted to Susan E. Coté, Susan J. Bakshi, Rickie Sue Cornfeld, Jane Kaho, Carrie Hendron Lester, Bernadine Psaty, Bruce Psaty and Richard Tomaszewski for their roles in the preparation of the manuscript.

Our thanks to Professors Harold S. Engelshon of Kingsborough Community College, Henry J. Ricardo of Manhattan College, and Donald S. Ostberg of Northern Illinois University.

*Cletus O. Oakley*
*Donald R. Kerr, Jr.*

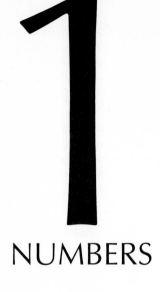

# NUMBERS

## 1.1    INTEGERS

You may find that you are familiar with many of the various topics
and much of the material in this chapter.   They are included both as
a review and as an introduction to elementary functions which is the
subject matter of the book as a whole.   Since functions involve relations among sets of numbers, we need to brush up on our knowledge of and experience with the numbers on the number line.   They
are called the *real numbers.*

It was in prehistory that people learned to count.   The first counting numbers, now called the *positive integers,* began with 1, 2, 3, . . . .
They did not have the number 0 (zero).   The notion of 0 was hard to
come by and is not much older than perhaps 3000 or 4000 years.   The
numbers 0, 1, 2, 3, . . . are called the *whole numbers.*   There is no
firm date for the introduction of the *negative integers,* $-1, -2, -3, \ldots,$
but it was well into the Christian Era.   The set of all of the integers,
positive, negative, and 0, is called simply the set of integers.   With

an appropriate choice of a unit length, we place these numbers on the number line as in Fig. 1.1.

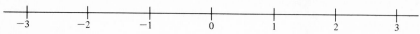

**Figure 1.1**

We have come to combine integers, as well as other numbers, using addition, subtraction, multiplication, and division. These operations reflect different mental and physical concepts. The operations of addition, subtraction, and multiplication of integers yield integers, whereas division may lead to nonintegers such as $\frac{2}{3}$, $\frac{5}{19}$, $-\frac{8}{7}$, and so forth.

## 1.2    UNIVERSE

Thus far we have spoken of the integers: They have been our universe of discourse. But we shall need other kinds of numbers, and these we shall treat in Secs. 1.3 to 1.7. Ultimately we shall want to use as our universe the set of numbers called the real numbers, which we shall describe shortly.

### DEFINITION 1.1

More specifically the *universe*, or universal set, is a predetermined, overall set or collection of elements of which sets, such as the integers, are considered subsets.  ◄

For example, in plane geometry (two dimensions), the universe of discourse is the set of all points in a plane. In solid geometry (three dimensions), the universe is the set of all points in space. At one time in your study of arithmetic, the set of whole numbers was considered the universal set. At another time it was the nonnegative rational numbers (fractions). In high school algebra, as well as in this course, the universe is usually the set of all the real numbers (the number line). In a given situation the universal set $U$ will either be understood and taken for granted, or it will be stated explicitly. In some instances we will not be interested in what the universe is.

## 1.3    VARIABLE

It is very likely that in your early school days you were asked questions like the following: What number does the square ☐ stand for in

$$\square + 2 = 5$$

if you want the statement to be true? A good teacher might also have asked you to select a number that makes the statement false. In algebra we are accustomed to the use of a letter of the alphabet, such as $x$, $y$, $z$, $a$, $b$, etc., instead of a square. Thus $x + 2 = 5$. Now as it stands, $x + 2 = 5$ is neither true nor false. We say, "It has no truth value." The $x$ is just a placeholder for a number and can be replaced by any number from our universe of numbers, whatever that universe may be. If we replace $x$ by 3, we get a true statement. If we replace $x$ by 4, we get a false statement.

### DEFINITION 1.2

A symbol, such as $x$, for which any element of the universe may be substituted, is called a *variable*. ◀

## 1.4   PROPOSITION

### DEFINITION 1.3

A proposition is a statement to which one and only one of the terms *true* or *false* can be meaningfully applied. ◀

Thus, "Today is Wednesday," is a proposition. We are writing this on Wednesday, so for us it is a true proposition. If you are reading it on Tuesday, then for you it is a false proposition. We shall be more interested in arithmetic statements, however.

### ▶ Illustration 1.1

Here are some examples of propositions involving numbers: $6 - 8 = -2$, $6 + 8 = 12$, $6 \times 8 = 14$, $\frac{6}{8} = \frac{3}{4}$, $\sqrt{16} = 4$, $-\sqrt{16} = 4$. Statements like these are all propositions, some of which are true and some of which are false. But each is a meaningful statement. ◀

## 1.5   OPEN SENTENCE

### DEFINITION 1.4

Any statement, containing a variable, which becomes a proposition when an element from the universe is substituted for the variable is called an *open sentence*. ◀

When values from the universe are substituted for the variables, the sentence becomes a proposition, true or false.

## DEFINITION 1.5

Those values that make the open sentence a true proposition are called *solutions*. ◀

The set of all numbers which are solutions is formally called the *solution set*, denoted by { }, but we prefer the more informal use of *solution(s)*.

### ▶ Illustration 1.2

The following is an open sentence: $x + 2 = 5$. It becomes a false proposition when we substitute 7 for the variable $x$. It becomes a true proposition when we substitute 3 for the variable, and the number 3 is the *solution* of the open sentence. ◀

Equivalent open sentences are those that have the same number(s) as solution(s).

### ▶ Illustration 1.3

For $x + 2 = 5$ some equivalent open sentences are $x = 5 - 2$, $7x + 14 = 35$, and $2 + \sqrt{9} + x = 8$. For each, the solution is the single number 3. ◀

The open sentence $x + 2 = 5$ is also called an *equation*. An equation may have several variables such as $x + 2y = 3$, $x^2 + y^2 = z$, etc. We say we have solved an equation when we have found all the numbers which, when substituted for the variable(s), turn the equation into a true statement in arithmetic.

## 1.6    RATIONAL NUMBERS

The solution of $x + 2 = 3$ is the positive integer 1. But the equation $2x + 2 = 5$ is surely satisfied by no integer. (If our universe were the integers, this equation would not have a solution.) To solve this equation we write a sequence of equivalent equations.

### ▶ Illustration 1.4

Let us start with the equation $2x + 2 = 5$.

Adding a constant c to both sides of such an equation will produce an equivalent equation. By adding $-2$ to both sides we get the equivalent equation $2x + 2 - 2 = 5 - 2$, or $2x = 3$.

Multiplying both sides of an equation by a constant $k \neq 0$, produces an equivalent equation. Multiplying by $\frac{1}{2}$ (that is, dividing by 2), we get another equivalent equation, $x = \frac{3}{2}$. The solution set is clearly $\{\frac{3}{2}\}$. However, we usually stop at the $x = \frac{3}{2}$ stage and say, "The solution is $x = \frac{3}{2}$," or simply, "The solution is the number $\frac{3}{2}$." ◀

The number $\frac{3}{2}$ belongs to the set of numbers called *rational numbers* according to the following definition.

## DEFINITION 1.6

A rational number is any number that can be expressed as the quotient of two integers. The quotient of two integers is often called a *fraction*. ◀

▶ **Illustration 1.5**

Some examples of rational numbers are $\frac{3}{7}$, $\frac{4}{15}$, $-\frac{6}{5}$, $\frac{0}{2}$ (=0), $\frac{5}{1}$ (=5), $\frac{8}{16}$, $7\sqrt{2}/8\sqrt{2}$ (=$\frac{7}{8}$), $\pi/3\pi$ (=$\frac{1}{3}$). Note that the rational numbers include the integers. For example, $\frac{10}{5}$ (rational) = 2 (integer), $\frac{90}{5} = 18$, etc. ◀

An important property of rational numbers is given in the following theorem.

---

**THEOREM 1.1**    *A rational number, when expressed decimally, is a repeating decimal.*

---

Before proving this theorem we give the following examples.

▶ **Illustration 1.6**

$$\frac{1}{2} = 0.5000 \cdots$$
$$\frac{2}{3} = 0.6666 \cdots$$
$$-\frac{9}{2} = -4.5000 \cdots$$
$$\frac{3}{11} = 0.272727 \cdots \quad ◀$$

---

**PROOF**

Consider the rational number $p/q$, where $p$ and $q$ are integers and $q \neq 0$. Also there is no loss in generality in assuming that the fraction $p/q$ is *reduced*, that is, $p$ and $q$ have no factor in common other than 1.

Now in dividing by $q$ no remainder, as the division process proceeds, can be as large as $q$, and each remainder is a nonnegative integer. Therefore the set of theoretically possible remainders is $\{0, 1, 2, \ldots, q-1\}$. The number of integers in this set is $q$, so that *after $q$ divisions at most* we must come up with a remainder that we have had before, and therefore the division process repeats. This proves the theorem.

◆ **Illustration 1.7**

Consider the rational $\frac{1}{7}$. The division process is indicated below in detail.

$$
\begin{array}{r}
0.1\ 4\ 2\ 8\ 5\ 7 \\
7\, \overline{\big)\ 1.0\ 0\ 0\ 0\ 0\ 0} \\
\text{Remainders} \quad 1\ 3\ 2\ 6\ 4\ 5\ 1
\end{array}
$$

We now have repeated the remainder 1, and so the division process repeats. Thus

$$\tfrac{1}{7} = 0.142857142857 \cdots$$

which we usually write

$$\tfrac{1}{7} = 0.\overline{142857}$$

where the bar indicates the digits that are repeated. ◆

◆ **Illustration 1.8**

Other examples are $\frac{1}{11} = 0.\overline{09}$, $\frac{1}{2} = 0.5\overline{0}$, $\frac{397}{1000} = 0.397\overline{0}$, and $\frac{1}{28} = 0.03\overline{571428}$. When 0 alone repeats, we usually drop it, as in $\frac{1}{2} = 0.5$, etc. ◆

The converse of Theorem 1.1 is also true, but we do not prove it:

---

**THEOREM 1.2**    *An infinite repeating decimal is a rational number.*

---

◆ **Illustration 1.9**

The following infinite repeating decimals are rational numbers.

$$0.\overline{1536}$$
$$-0.297\overline{504}$$
$$8.\overline{1}$$
$$23.\overline{0} \quad \blacktriangleleft$$

So there are two ways of representing rational numbers, one using fractions, the other using repeating decimals such as $1.\overline{6}$, $0.5\overline{0}$, $7.\overline{652}$, etc.  It is useful to switch back and forth between these representations.  A repeating decimal can be converted to a fraction as in the illustrations below.

**Illustration 1.10**

Find a rational number $x$ (in the form of $p/q$) such that $x = 0.\overline{37}$.  Multiplying by 100, we get

$$100x = 37.\overline{37} = 37 + 0.\overline{37} = 37 + x$$

from which it follows that $100x - x = 37$ and

$$99x = 37$$
and $$x = \tfrac{37}{99} \quad \blacktriangleleft$$

**Illustration 1.11**

Find a rational number $x$ (in the form $p/q$) such that $x = 0.\overline{456}$.  Multiplying by 1000 we get

$$1000x = 456.\overline{456} = 456 + 0.\overline{456} = 456 + x$$
Hence      $$1000x - x = 456$$
and      $$999x = 456$$
and      $$x = \tfrac{456}{999}$$

(We do not need to reduce this to $\tfrac{152}{333}$.)  $\blacktriangleleft$

**Illustration 1.12**

Express $0.\overline{1258}$ as a fraction.  We multiply by 10,000, getting

$$10{,}000x = 1258 + 0.\overline{1258} = 1258 + x$$
$$9999x = 1258$$
and      $$x = \tfrac{1258}{9999} \quad \blacktriangleleft$$

Do you see a pattern? Now write down immediately the repeating decimal $0.\overline{22345}$ as the quotient of two integers. Did you get 22,345/99,999? Try $x = 2.\overline{37}$ to find $x = \frac{235}{99}$. Try $x = 0.5\overline{7}$. Write $10x = 5 + 0.\overline{7} = 5 + \frac{7}{9} = \frac{52}{9}$. Hence $x = \frac{52}{90}$.

These techniques depend heavily on the fact that the decimal is repeating. (Do you see why?) What can one do with a nonrepeating decimal such as $0.12112111211112\cdots$? The answer is that one cannot represent such a number as a fraction. Such nonrepeating decimals are called irrational numbers, and we say a few words about them in the next section.

The sum, difference, product, and quotient of any two rational numbers are rational numbers. (We do not divide by 0. See Sec. 1.12.)

♦ **Illustration 1.13**

Sum:
$$2 + \tfrac{5}{8} = \tfrac{21}{8}$$
$$3 + (-7) = -\tfrac{4}{1}$$

Difference:
$$6 - \tfrac{3}{2} = \tfrac{9}{2}$$
$$4 - (-3) = \tfrac{7}{1}$$

Product:
$$\tfrac{2}{3} \times \tfrac{7}{5} = \tfrac{14}{15}$$
$$-\tfrac{6}{5} \times \tfrac{2}{11} = -\tfrac{12}{55}$$

Quotient:
$$\frac{\tfrac{2}{3}}{\tfrac{7}{11}}$$

To simplify this we multiply the numerator *and* the denominator by $\frac{11}{7}$. You will see why as we do it.

$$\frac{\tfrac{2}{3} \times \tfrac{11}{7}}{\tfrac{7}{11} \times \tfrac{11}{7}}$$

$$\frac{\tfrac{2}{3}}{\tfrac{7}{11}} = \frac{\tfrac{2}{3} \times \tfrac{11}{7}}{\tfrac{7}{11} \times \tfrac{11}{7}}$$

Now the new denominator is 1, since $\frac{7}{11} \times \frac{11}{7} = \frac{77}{77} = 1$.

Thus
$$\frac{\tfrac{2}{3}}{\tfrac{7}{11}} = \frac{2}{3} \times \frac{11}{7}$$

and this reduces to the rational fraction $\frac{22}{21}$.

$$\frac{-\tfrac{4}{13}}{\tfrac{15}{7}} = -\frac{4}{13} \times \frac{7}{15} = -\frac{28}{195} \quad ♦$$

The rules governing the examples just given are summarized as follows:

If the rational numbers are expressed in fractional form, $a/b$ and $c/d$, where neither $b$ nor $d$ is zero, then their sum can be expressed as follows (first reducing each fraction to a common denominator):

$$\frac{a}{b} + \frac{c}{d} = \frac{ad}{bd} + \frac{cb}{db} = \frac{ad + bc}{bd}$$

Similarly

$$\frac{a}{b} - \frac{c}{d} = \frac{ad}{bd} - \frac{cb}{db} = \frac{ad - bc}{bd}$$

Since $a$, $b$, $c$, and $d$ are integers, it follows that $ad$, $bc$, $bd$, and $ad + bc$ are integers, so the sum of two rationals is a rational. The product of $a/b$ and $c/d$ is

$$\frac{a}{b} \times \frac{c}{d} = \frac{ac}{bd}$$

which is rational.

The quotient

$$\frac{\dfrac{a}{b}}{\dfrac{c}{d}}$$

can be written in simpler form by multiplying numerator and denominator by $d/c$. (Recall that this is multiplying the quotient

$$\frac{\dfrac{a}{b}}{\dfrac{c}{d}}$$

by 1.) We get

$$\frac{\dfrac{a}{b}}{\dfrac{c}{d}} = \frac{\dfrac{a}{b} \times \dfrac{d}{c}}{\dfrac{c}{d} \times \dfrac{d}{c}} = \frac{a}{b} \times \frac{d}{c} = \frac{ad}{bc} \qquad c \neq 0$$

and $ad/bc$ is a rational number.

How are the rational numbers spread over the number line? Are there any gaps? To see the answer we look at the arithmetic mean or average of $a/b$ and $c/d$, defined to be one-half of the sum,

$$\frac{\dfrac{a}{b} + \dfrac{c}{d}}{2}$$

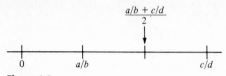

**Figure 1.2**

If $a/b < c/d$, then these numbers look like Fig. 1.2 on the number line. ($a/b < c/d$ is read "$a/b$ is less than $c/d$.")  The average is midway between the numbers $a/b$ and $c/d$.  This is easy to prove, and the proof is called for in the problems.  The import is enormous.  This means that in between any two rationals there is another rational, namely, the average of the two: $(r_1 + r_2)/2$ is rational and in between $r_1$ and $r_2$. And this implies that in between any two rationals there is an infinity of rationals, so that on the number line the rationals are pretty thickly spread.  The integers, being isolated, do not have this property.

## PROBLEMS 1.6

**1**  Draw a number line similar to Fig. 1.1 and place the following numbers on it at their appropriate points.

    **a** $-\frac{2}{3}$    **b** $\frac{3}{7}$    **c** $\frac{4}{9}$    **d** 2.25

**2**  Draw a number line similar to Fig. 1.1 and place the following numbers on it at their appropriate points.

    **a** $\frac{3}{2}$    **b** 0.75    **c** $-\frac{4}{7}$    **d** $-\frac{5}{9}$

**3**  Find all numbers which satisfy

    **a** $3x - 4 = 8$    **b** $4x - 2 = \frac{3}{4}$    **c** $5x = 0$    **d** $x = x + 1$

**4**  Find all numbers which satisfy

    **a** $4 - 5x = 14$    **b** $\frac{2}{3}x + 2 = 4$

    **c** $x + \frac{3}{2} = -x - \frac{3}{2}$    **d** $4x = 4x$

**5**  Write the following numbers in repeating-decimal form.

    **a** $\frac{2}{13}$    **b** $\frac{5}{13}$    **c** $\frac{2}{39}$    **d** $\frac{7}{39}$

**6**  Write the following numbers in repeating-decimal form.

    **a** $\frac{7}{13}$    **b** $\frac{9}{13}$    **c** $\frac{1}{44}$    **d** $\frac{1}{16}$

**7**  Write the following repeating decimals as rational numbers in fraction form.

    **a** $0.\overline{549}$    **b** $0.0\overline{15}$    **c** $0.\overline{027}$    **d** $1.\overline{81}$

**8** Write the following repeating decimals as rational numbers in fraction form.

    **a** $0.\overline{456}$     **b** $0.0\overline{18}$     **c** $0.\overline{0099}$     **d** $3.\overline{27}$

**9** Prove that the point on the number line which represents the arithmetic mean of two given rationals lies midway between the points representing the two rationals. Hint: Make use of the fact that the length of a segment on the number line from point $p$ to point $q$ (where $p < q$) is $q - p$.

## 1.7   IRRATIONAL NUMBERS

We have mentioned that the rational numbers are spread thickly on the number line. Yet, strange as it may seem, they do not fill up the number line — there are still a lot of blank spots (points) on the number line *not* occupied by a rational number. These points are available for the representation of other kinds of numbers such as $\sqrt{2}, \sqrt{3}, \sqrt{19}, -2\sqrt{2}, \pi, -2\pi, \sqrt{2\pi}, 37^{2/3}, \log_{10} 7, \sin 16°$. In fact there are more of these, called *irrational numbers*, than there are rational numbers. An irrational number, when written in decimal form, is an infinite *nonrepeating* decimal. We can write a rational number approximation to an irrational for numerical work. Thus, correct to five decimal places, $\pi = 3.14159$, $\sqrt{2} = 1.41421$, etc.

The rationals and the irrationals make up all of the *real* numbers and, together, they occupy all the points on the number line. When we say "together," we mean of course the *union* of the two sets — a concept familiar to most elementary school children. Since both union and *intersection* will be used in Chap. 2, now is a good time for a bit of review. We begin with some examples.

The union of the two sets $\{\pi, a, \$\}$ and $\{6, F, *, \sqrt{\ }\}$ is the set $\{\pi, a, \$, 6, F, *, \sqrt{\ }\}$. This is written

$$\{\pi, a, \$\} \cup \{6, F, *, \sqrt{\ }\} = \{\pi, a, \$, 6, F, *, \sqrt{\ }\}$$

The union of $\{\pi, a, \$\}$ and $\{6, \pi, F, \$\}$ is

$$\{\pi, a, \$\} \cup \{6, \pi, F, \$\} = \{\pi, a, \$, 6, F\}$$

We do not duplicate an element which appears in both sets. In this example $\pi$ and $\$$ are in each set. In the union these elements are not written twice.

### DEFINITION 1.7

The union of two sets is the set containing once and only once each element found in either or both of the given sets. ◆

◆ **Illustration 1.14**

The union of the fractions and the integers is the fractions (rational numbers).

$$\{\text{Rationals}\} \cup \{\text{irrationals}\} = \{\text{reals}\} \quad ◀$$

The intersection of two sets $A$ and $B$ is written $A \cap B$.

$$\{\pi, a, \$\} \cap \{6, \pi, F, \$\} = \{\pi, \$\}.$$

In taking the intersection of two sets we take all elements, and only those elements that are common to the two sets. Here the common elements are $\pi$ and $\$$.

**DEFINITION 1.8**

The intersection of two sets is the set containing once and only once the elements common to the two given sets. ◀

◆ **Illustration 1.15**

$$\{\text{Rationals}\} \cap \{\text{integers}\} = \{\text{integers}\}$$
$$\{\text{Reals}\} \cap \{\text{positive integers}\} = \{\text{positive integers}\} \quad ◀$$

Now how about taking the intersection of the rationals and the irrationals? Since these two sets are disjoint, that is, they have no element in common, we are led to the notion of a *set with no elements in it!* It turns out to be a useful concept. The set containing no elements is called the *null set,* the symbol for which is either { } or∅. Do not confuse the null set with the set containing only the number 0. This is written $\{0\}$.

Throughout this book we shall be using the word interval. We need the following definitions.

**DEFINITIONS 1.9**

An *interval* on the number line is the set of all points between $a$ and $b$ including (or excluding) either $a$ or $b$ or both.

The interval is closed if the endpoints $a$ and $b$ are included.

The interval is open if $a$ and $b$ are excluded.

The interval is open at $a$ and closed at $b$ if $a$ is excluded and $b$ is included.

An interval consisting of all points on one side of point $a$ is usually called a *half-line,* and it may be open or closed at $a$. ◀

There are various notations for an interval. For example, the interval from −1 to 4, inclusive of −1 and 4, is often written in the form of a double inequality, $-1 \leq x \leq 4$, which is read "the interval from −1 to 4 inclusive of the endpoints." Or, again, in so-called *set-builder* notation, we write $\{x \mid -1 \leq x \leq 4\}$ which is read "the set of all x such that −1 is less than or equal to x which, in turn, is less than or equal to 4." It can also be read "the set of all x such that x is greater than or equal to −1 and less than or equal to 4."

▶ **Illustration 1.16**

Other examples are:

| INTERVAL | SET-BUILDER NOTATION |
|---|---|
| $6 < x \leq 7$, open at 6 | $\{x \mid 6 < x \leq 7\}$ |
| $-10 < x < -3$, open | $\{x \mid -10 < x < -3\}$ |
| $0 \leq x < 9$, open at 9 | $\{x \mid 0 \leq x < 9\}$ |

| HALF-LINE | SET-BUILDER NOTATION |
|---|---|
| $-5 \leq x$ | $\{x \mid -5 \leq x\}$ |
| $x < 2$ | $\{x \mid x < 2\}$  ◀ |

**1.8    COMPLEX NUMBERS**

Since the real numbers will be our universe of discourse, one of our objectives will be to find all real numbers that satisfy a given equation or inequality. But right here we should point out that, while a given equation might have one or more real numbers as solutions, it is possible for an equation to have "numbers" other than real numbers as solutions. Consider, the equation $x^2 + 9 = 0$. Zero is not a solution. In fact, no real number satisfies this equation. We can see this quickly by noting that regardless of what real number, positive or negative, we substitute for x, the square of it is positive and the sum of two positive numbers is positive, and so the equation cannot be satisfied by any real number. Now let us formally solve this equation by the usual means. We write

$$x^2 + 9 = 0 \qquad \text{or} \qquad x^2 = -9$$

and
$$x = \pm\sqrt{-9}$$

So there are two solutions of the equation, neither of which is a real number. Indeed to make sense, and "numbers," out of $\pm\sqrt{-9}$, we must make definitions, lay down rules of operation, etc. Such numbers are members of the set of *complex* numbers; we shall not use them in this book.

## 1.9    ABSOLUTE VALUE

We now need to pick up a new idea and a new notation. A real number actually has two qualities: It has *magnitude* (how big or small is it?) and *direction* (is it *positive* and to the *right* of the origin or is it *negative* and to the *left* of the origin?). Sometimes we are interested only in magnitude and not in whether the number lies to the right or to the left of the origin. We say that the absolute value of 7, written $|7|$, is just 7; the number is at a distance of 7 units from the origin — to the right. We also say the absolute value of $-7$, written $|-7|$, is 7 and that this number 7 is 7 units from the origin, also to the right. That is, $|7| = |-7|$.

◆ **Illustration 1.17**

Other examples are: $|-3| = 3$, $|\frac{1}{2}| = \frac{1}{2}$, $|0| = 0$. ◆

**DEFINITION 1.10**

The *absolute value* of $a$, written $|a|$, is

$$|a| = \begin{cases} a & \text{if } a \geq 0 \\ -a & \text{if } a < 0 \end{cases}$$

where $a$ is a real number. ◆

◆ **Illustration 1.18**

$$
\begin{aligned}
&|9| = 9 && \text{since } 9 > 0 \\
&|-13| = 13 && \text{since } -13 < 0 \text{ and } -(-13) = 13 \\
&|-1| = 1 && \text{since } -1 < 0 \text{ and } -(-1) - 1 \\
&|8| = 8 && \text{since } 8 > 0 \\
&|x^2| = x^2 && \text{since } x^2 \geq 0 \\
&|-x^2| = x^2 && \text{since } -x^2 < 0 \text{ and } -(-x^2) = x^2 \\
&|x - 3| = \begin{cases} x - 3 & \text{if } x - 3 \geq 0 \\ -(x-3) & \text{if } x - 3 < 0 \end{cases} &&
\end{aligned}
$$
◆

▶ **Illustration 1.19**

Find all numbers satisfying the equation $|x - 5| = 7$.

*Case 1*   If $x - 5 \geq 0$, then $|x - 5| = x - 5$ and the equation reads $x - 5 = 7$. Hence $x = 12$.

*Case 2*   If $x - 5 < 0$, then $|x - 5| = -(x - 5)$ and so $|x - 5| = -x + 5$, and the equation is $-x + 5 = 7$.   Hence $x = -2$.
Two and only two numbers, namely, 12 and $-2$, satisfy the given equation.

*Check, Case 1*            $|12 - 5| = 7$      O.K.

*Check, Case 2*            $|-2 - 5| = 7$      O.K.  ◀

▶ **Illustration 1.20**

Solve the equation $|-x + 2| = 2x + 1$.

*Case 1*   If $-x + 2 \geq 0$, then $-x + 2 = 2x + 1$, and $2 - 1 = 2x + x$, or $x = \frac{1}{3}$.

*Case 2*   If $-x + 2 < 0$, then $-(-x + 2) = 2x + 1$, or $x - 2 = 2x + 1$, and $-2 - 1 = 2x - x$, or $x = -3$.

*Check, Case 1*            $|-\frac{1}{3} + 2| \stackrel{?}{=} 2(\frac{1}{3}) + 1$

$\frac{5}{3} = \frac{5}{3}$     O.K.

*Check, Case 2*            $|-(-3) + 2| \stackrel{?}{=} 2(-3) + 1$

$|5| \stackrel{?}{=} -5$     No!

Case 2 fails and we could have seen this by noting that if $x = -3$, then it is false that $-x + 2 < 0$.   As a matter of fact $-(-3) + 2 = 5$ and 5 is not less than 0.   The only solution is $x = \frac{1}{3}$.  ◀

You should play around with the absolute-value notion until you can handle it.   We make considerable use of it in Chap. 2.

**1.10   ROOTS AND POWERS**

At least some of the following special notations relating to roots and powers should be familiar to you.   The product of $n$ $a$'s is simply written as follows

$$a \times a \times a \times \cdots \times a = a^n$$

▶ **Illustration 1.21**

$$a \times a \times a = a^3$$
$$5 \times 5 \times 5 \times 5 = 5^4$$

and $$4 \times 4 = 4^2$$

We call $1/a^n$ the reciprocal of $a^n$ and write

$$\frac{1}{a^n} = a^{-n} \quad ◀$$

▶ **Illustration 1.22**

$$\frac{1}{7^2} = 7^{-2}$$

$$\frac{1}{6^3} = 6^{-3}$$

and $$\frac{1}{10^{-4}} = 10^4$$

From the above it follows that

$$a^m \times a^n = a^{m+n} \quad \text{and} \quad \frac{a^m}{a^n} = a^{m-n}$$

where $m$ and $n$ are integers, and since

$$1 = \frac{a^m}{a^m} = a^{m-m}$$

we define $a^0$ to be 1; thus $a^0 = 1$. ◀

▶ **Illustration 1.23**

$$6^2 \times 6^3 = (6 \times 6) \times (6 \times 6 \times 6) = 6^{2+3} = 6^5$$
$$(\tfrac{1}{9})^3 \times (\tfrac{1}{9})^7 = (\tfrac{1}{9})^{10}$$
$$\frac{3^4}{3^2} = \frac{3 \times 3 \times 3 \times 3}{3 \times 3} = 3^{4-2} = 3^2$$
$$\frac{5^2}{5^2} = 5^0 = 1$$

and $$\frac{4^5}{4^8} = 4^{5-8} = 4^{-3}$$

It is also true that

$$(ab)^m = a^m b^m \qquad \text{and} \qquad (a^m)^n = a^{mn} \quad \blacktriangleleft$$

▶ **Illustration 1.24**

$$(2 \times 3)^4 = 2^4 \times 3^4$$
$$(5^2)^3 = 5^2 \times 5^2 \times 5^2 = 5^{2 \times 3} = 5^6$$

The numbers $m$ and $n$ do not have to be integers. For example, note that, for $a$ greater than or equal to 0 (written $a \geq 0$), it is true that

$$\sqrt{a} \times \sqrt{a} = a = a^1$$

It is therefore reasonable to write

$$\sqrt{a} = a^{1/2}$$

read "$a$ to the one-half power,"

since $\qquad\qquad a^{1/2} \times a^{1/2} = a^{1/2+1/2} = a^1$

Thus $\qquad\qquad 3^{1/2} \times 3^{1/2} = 3^{1/2+1/2} = 3^1 = 3$

hence $\qquad\qquad 3^{1/2} = \sqrt{3}$

Similarly $\qquad 5^{1/3} \times 5^{1/3} \times 5^{1/3} = 5^{1/3+1/3+1/3} = 5^{3/3} = 5$

and therefore $\qquad\qquad 5^{1/3} = \sqrt[3]{5} \quad \blacktriangleleft$

▶ **Illustration 1.25**

$$6 \times 6 \times 6 \times 6 \times 6 = 6^5$$
$$\frac{1}{9 \times 9 \times 9} = \frac{1}{9^3} = 9^{-3}$$
$$16 \times 16 \times 16 = 2^4 \times 2^4 \times 2^4 = 2^{12}$$
$$81 \times 25 = 3^4 \times 5^2$$
$$(11 \times 17)^2 = 11^2 \times 17^2$$
$$(15^3)^2 = 15^6$$
$$\sqrt[4]{7} = 7^{1/4} \qquad \sqrt[4]{4} = 4^{1/4} = (4^{1/2})^{1/2} = 2^{1/2}$$
$$\sqrt[3]{5^2} = (5^2)^{1/3} = 25^{1/3} \qquad \sqrt[2]{5^3} = 5^{3/2} \quad \blacktriangleleft$$

We shall explore these ideas in more generality in Sec. 6.3, but for the present we shall not elaborate. The following four numbers are the same number, written in different but equivalent notations.

$$\sqrt[7]{6^3} = 6^{3/7} = (6^{1/7})^3 = (6^3)^{1/7}$$

In general, with $q \neq 0$,

$$a^{p/q} = (a^{1/q})^p = \sqrt[q]{a^p} = (a^p)^{1/q}$$

♦ **Illustration 1.26**

$$\sqrt[3]{2^5} = (2^{1/3})^5 = (2^5)^{1/3} = (32)^{1/3} = \sqrt[3]{32}$$

and we can simplify this a bit by writing

$$\sqrt[3]{32} = \sqrt[3]{8 \times 4} = \sqrt[3]{8} \; \sqrt[3]{4}$$
$$= 2\sqrt[3]{4} = 2(1.5874) \qquad \text{approximately, by calculator}$$
$$= 3.1748 \quad ♦$$

Odd roots of negative numbers are negative. For example $\sqrt[3]{-8} = -2$. Where roots and powers are present, numerical computations can be done by logarithms, by other tables, or by calculators.

## 1.11    RULES OF ARITHMETIC

In the previous section a single real number, say $a$, was raised to some rational number power, say $p/q$, $q \neq 0$. Such an operation is called a *unary* operation since it operates on a single number. Addition is another arithmetic operation but it is a *binary* operation. You do not add *one* number — you add two numbers and just two numbers — you *never* add three or more.

Add $7 + 4 + 9$. What do you do? You add, first, say 7 and 4. The sum is a new number 11. Then you add 11 and 9. Your operations are binary: You add two and only two numbers at a time.

Now note that subtraction is not really involved: We *add* real numbers, whether they be positive or negative.

♦ **Illustration 1.27**

$$6 + 4 = 10 \qquad 9 + (-3) = 6$$

Sometimes we *call* this subtraction and write $9 - 3 = 6$, but it is really addition of 9 and $-3$.

Likewise multiplication is a binary operation:

$$3 \times 7 = 21$$
$$4 \times (-6) = -24$$
$$7 \times \tfrac{1}{9} = \tfrac{7}{9}$$

This looks like something we call division. But, again, it is really just multiplication of 7 by $\tfrac{1}{9}$. ♦

Subtraction and division have been introduced to express addition and multiplication by inverses in order to simplify notation and language. They also facilitate the relationship with the "take away" and "sharing" models used with children.

We now give a batch of numerical examples indicating that, in most cases, simplification is possible.

▶ **Illustration 1.28**

*Addition:*

$$2 + \tfrac{7}{3} = \tfrac{13}{3} \qquad\qquad\text{rational}$$

$$\frac{4}{5} + \left(-\frac{3}{7}\right) = \frac{4}{5} - \frac{3}{7} = \frac{28 - 15}{35} = \frac{13}{35} \qquad\text{rational}$$

$$\sqrt{2} + \sqrt{2} = 2\sqrt{2} \qquad\qquad\text{irrational}$$

$$\sqrt{2} + \sqrt{3} = \sqrt{2} + \sqrt{3} \qquad\qquad\text{irrational (cannot simplify in this form)}$$

$$\sqrt{2} + (-\sqrt{2}) = \sqrt{2} - \sqrt{2} = 0 \qquad\text{integer} \ \blacktriangleleft$$

▶ **Illustration 1.29**

*Multiplication:*

$$4 \times \tfrac{3}{7} = \tfrac{12}{7} \qquad\qquad\qquad\text{rational}$$

$$(-4) \times \sqrt{7} = -4\sqrt{7} \qquad\qquad\text{irrational}$$

$$(2 + \sqrt{3}) \times (4 - 6\sqrt{11}) = (2 \times 4) - (2)6\sqrt{11}$$
$$+ 4\sqrt{3} - 6\sqrt{11}\,\sqrt{3}$$
$$= 8 - 12\sqrt{11} + 4\sqrt{3} - 6\sqrt{33} \qquad\text{irrational}$$

$$(\sqrt{5}) \times \frac{2}{3} = \frac{2\sqrt{5}}{3} \qquad\qquad\qquad\text{irrational} \ \blacktriangleleft$$

## PROBLEMS 1.11

**1** Given the two rational numbers $p$ and $q$, show that the quotient $p/q$ is a rational number.

    **a** $p = \tfrac{2}{3}, q = \tfrac{8}{7}$      **b** $p = -\tfrac{9}{4}, q = \tfrac{3}{2}$

    **c** $p = \tfrac{5}{6}, q = -\tfrac{6}{5}$      **d** $p = -\tfrac{1}{2}, q = -\tfrac{10}{11}$

**2** Given the two rational numbers $p$ and $q$, show that the quotient $p/q$ is a rational number.

    **a** $p = -\tfrac{4}{9}, q = \tfrac{8}{9}$      **b** $p = \tfrac{10}{3}, q = -\tfrac{10}{3}$

    **c** $p = \tfrac{15}{16}, q = 5$      **d** $p = \tfrac{2}{3}, q = \tfrac{2}{4}$

**3**  **a**  Is the sum of two irrational numbers always an irrational number? Prove your answer.

**b**  Is the product of two irrational numbers always an irrational number?  Prove your answer.

**4**  **a**  Is the difference of two irrational numbers always an irrational number?  Prove your answer.

**b**  Is the quotient of two irrational numbers always an irrational number?  Prove your answer.

**5**  Find the product in each case.

**a**  $0.87 \times 2$          **b**  $0.8787 \times 2$

**c**  $0.878787 \times 2$     **d**  $0.\overline{87} \times 2$

**6**  Find the product in each case.

**a**  $0.64 \times 3$          **b**  $0.6464 \times 3$

**c**  $0.646464 \times 3$     **d**  $0.\overline{64} \times 3$

**7**  Write $0.37\overline{425}$ as a fraction.  Hint: Let $x = 0.37\overline{425}$, $100x = 37 + 0.\overline{425}$. First find the fraction representing $0.\overline{425}$.

**8**  Write $2.5\overline{37}$ as a fraction.  Hint: Let $x = 2.5\overline{37}$, $10x = 25 + 0.\overline{37}$.  First find the fraction representing $0.\overline{37}$.

**9**  Find all numbers which satisfy each equation.

**a**  $|x| = 4$          **b**  $|x| = -4$

**c**  $|x - 3| = 4$.     Hint: Case 1.  If $x - 3 > 0$, then $|x - 3| = x - 3$. Case 2.  If $x - 3 < 0$, then $|x - 3| = -(x - 3)$.  Review the definition of absolute value, Sec. 1.9.

**d**  $|x + 3| = 4$.  See part c above and paraphrase.

**e**  $|x - 3| = x + 4$.  See part c above.

**f**  $|x^2 - 2| = 1$.  See part c above.

**10**  Find all numbers which satisfy each equation.

**a**  $|-x| = 3$          **b**  $|-x| = -3$

**c**  $|4 - x| = 3$.     Hint: Case 1.  If $4 - x > 0$, then $|4 - x| = 4 - x$. Case 2.  If $4 - x < 0$, then $|4 - x| = -4 + x$.  Review the definition of absolute value, Sec. 1.9.

**d**  $|-x - 6| = 2$       **e**  $|x + 4| = |3x|$       **f**  $|x^2 + 1| = 2$

**11**  Simplify using only positive powers of 2 and 3.

**a**  $\dfrac{2^{-3} \cdot 3^5}{27 \cdot 8}$     **b**  $\dfrac{6 \cdot 12}{2^{-3} \cdot 3^{-2}}$     **c**  $\dfrac{4.6^{-3} \cdot 8^{1/2}}{\sqrt[3]{12}}$     **d**  $\dfrac{24 \cdot 32^{1/2}}{6^{-2} \cdot 2^{-3}}$

**12**  Simplify using only positive powers of 2, 3, and 5.

**a**  $\dfrac{25 \cdot 12^2}{30}$          **b**  $\dfrac{40 \cdot 16}{90 \cdot 2^8 \cdot 5^{-1}}$

$$\text{c} \quad \frac{60 \cdot \sqrt{3} \cdot 5^{3/2}}{48 \cdot 5^4} \qquad \text{d} \quad \frac{2^{-3} \cdot 9^{-4} \cdot 25^{3/2}}{\sqrt{5} \cdot 4 \cdot 3}$$

**13**   In each case write the quotient in the form $A + B\sqrt{C}$, where $A$ and $B$ are rational numbers and $C$ is a positive integer.  (The process in the hints is called *rationalizing the denominator*.)

$$\text{a} \quad \frac{2 - \sqrt{5}}{3 - \sqrt{5}} \quad \text{Hint: Multiply numerator and denominator by } 3 + \sqrt{5}$$
and simplify.

$$\text{b} \quad \frac{4 - \sqrt{3}}{2 + \sqrt{3}} \quad \text{Hint: Multiply numerator and denominator by } 2 - \sqrt{3}$$
and simplify.

$$\text{c} \quad \frac{7 + 2\sqrt{2}}{1 - 4\sqrt{2}}$$

$$\text{d} \quad \frac{4 + \sqrt{6}}{3 + \sqrt{6}}$$

## 1.12   ZERO

We close this chapter with a few remarks on some special properties of the special integer 0.

*Property 1*   It is the only integer without a sign: 0 is neither positive nor negative, although we do write $a + 0$ (for addition) and $a - 0$ (for subtraction).

*Property 2*   Zero is the only number such that, for every real number $a$ it is true that

$$a + 0 = 0 + a = a$$

The sum is just $a$ again and 0 is called the *additive identity*, since $a$ retains its identity under addition with 0.

For the next properties we make use of the fact that $a/b = c$ and $a = bc$ are equivalent statements, provided $b \neq 0$.

*Property 3*   For any real number $a$, not 0,

$$\frac{0}{a} = 0$$

The truth of this statement becomes immediately obvious if we write it in the form $0 = a \times 0$, since $a \times 0 = 0$ regardless of what real number $a$ is.

*Property 4*   For any real number $a$,

$$\frac{a}{0} \text{ is meaningless} \qquad a \neq 0$$

For suppose $a/0 = c$, where $c$ is a real number. Now we write this in the equivalent form $a = 0 \times c$. But $a$ is not 0 and therefore cannot be equal to $0 \times c$ which is 0. This proves that $a/0$ is not a number, and we say that $a/0$ does not exist.

*Property 5*   This might be called the "powerhouse" property, because it arises in the theory of limits, which is the very essence of the calculus and the vast branch of mathematics called analysis:

$$\frac{0}{0} \text{ is undefined}$$

Let us suppose that $0/0 = c$, where $c$ is a real number. Then, writing this in the equivalent form $0 = 0 \times c$, we note that this is true for each and every real number $c$ without exception. So the ratio $0/0$ is simply not defined.

Properties 4 and 5 say that *we never divide by 0.*

**PROBLEMS 1.12**

   **1**   If $x = 0$, then $x(x - 1) = 0$ and $x - 1 = 0$, or $x = 1$, from which it follows that $0 = 1$. What is wrong with the work?
   **2**   If $x = 2$, then on multiplying by $x - 1$ we get $x^2 - x = 2x - 2$. Now we subtract $x$ from both sides and get $x^2 - 2x = x - 2$, or $x(x - 2) = x - 2$. Divide by $x - 2$. It follows that $x = 1$, so $2 = 1$. What is wrong with the work?

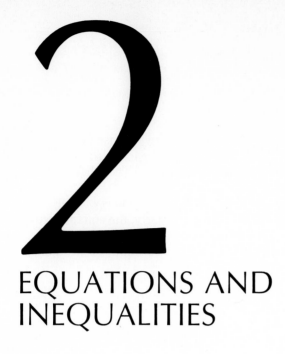

# EQUATIONS AND INEQUALITIES

## 2.1 IDENTITIES AND EQUATIONS

The simple notions of an open sentence or equation and equivalent equations were considered in Chap. 1 and with them we solved some equations. In this chapter we shall solve more equations and also some inequalities.

Some equations involving a variable are always true, some are never true, and some are true for some values of the variable and false for other values.

▶ **Illustration 2.1**

For example, the equation $2(x + 1) = 2x + 2$ is true for all real numbers $x$, and the equation $1/3(x - 1) = 1/(3x - 3)$ is true for all real numbers $x$ except $x = 1$ where the equation is not defined since division by 0 is not defined. ◀

## DEFINITION 2.1

Equations that are true for all permissible values of the variable are called *identities*. ◂

The equation $x = x + 1$ is never true because no number can be 1 more than itself, and the equation $x^2 = -1$ is also never true for real numbers because the square of a real number is either positive or 0 and obviously 0 is not a solution.

Finally, the equation $x + 2 = 5$ is true when $x$ has the value 3 and false when $x$ has any other value. Similarly, the equation $x^2 = 1$ is true when $x$ has the value 1 and also when $x$ has the value $-1$, but it is false when $x$ has any other value. Thus the solutions of $x^2 = 1$ are 1 and $-1$.

◂ **Illustration 2.2**

Here are some identities:

$$x + 2 = x + 2 \qquad\qquad 2(x - 7) = 2x - 14$$
$$(x + 2)^2 = x^2 + 4x + 4 \qquad x(2x + 8) = 2x^2 + 8x$$
$$(x + 2)^3 = x^3 + 6x^2 + 12x + 8 \qquad \sqrt{x} = \sqrt{x} \qquad x \text{ nonnegative}$$
$$x - 9 = x - 9 \qquad\qquad x/x = 1, \ x \neq 0$$
$$|x - 11| = |-x + 11| \qquad \sqrt{x^2} = |x| \ ◂$$

We make use of algebraic identities from time to time, but we shall not dwell on them here. In Chap. 9 there will be many trigonometric identities to study.

## 2.2    FIRST-DEGREE EQUATIONS

About the simplest equation in one variable is one in which the variable, say $x$, appears only to the first power, namely $ax + b = 0$ (or an equation which can be reduced to this form), where $a$, $b$ are real numbers and $a \neq 0$. Such an equation is called a *first-degree* or linear equation. It is very easily solved $ax = -b$ or $x = -b/a$. Here are some numerical examples.

| EQUIVALENT OPEN SENTENCES | SOLUTION SET |
|---|---|
| **a**   $2x - 8 = 0, \ 2x = 8, \ x = 4$ | $\{4\}$ |
| **b**   $\sqrt{7}x + 8 = 0, \ x = -8/\sqrt{7}$ | $\{-8/\sqrt{7}\}$ |

**c**   $5x + 10 = 0, x = -2$          $\{-2\}$

**d**   $9x - 27 = 0, x = 3$          $\{3\}$

**e**   $2x - 5 = x + 6, 2x - x - 5 - 6 = 0,$

   $x - 11 = 0, x = 11$          $\{11\}$

**f**   $x + 2(3 - x) = 5, -x + 6 = 5, x = 1$   $\{1\}$

**g**   $x + 7 = x - 8$          $\{\ \}$

On the number line, the solution of a first-degree equation is one point. In g above, note that the equation is not of first degree.

**2.3     QUADRATIC EQUATIONS**

A quadratic equation has the form

$$ax^2 + bx + c = 0$$

where $a$, $b$, and $c$ are real numbers and $a \neq 0$. Keep in mind that $a$ and $b$ are the coefficients of $x^2$ and $x$, respectively, and $c$ is the constant term. This equation is not terribly hard to solve — once you see how the method works. The Rhind papyrus (Egyptian, 1700 B.C.) now in the British Museum, shows clearly that the mathematicians of that era knew how to solve the quadratic equation.

Before doing the algebra for the general quadratic with coefficients $a$, $b$, and $c$, we consider a numerical example.

▶ **Illustration 2.3**

We want the numbers that satisfy

$$2x^2 + x - 1 = 0 \qquad a = 2, b = 1, \text{ and } c = -1 \qquad (1)$$

*Step 1*   Add $+1$ to both sides:

$$2x^2 + x = 1$$

(This is equivalent to transposing the constant $-1$.)

*Step 2*   Divide by the coefficient of $x^2$:

$$x^2 + \frac{x}{2} = \frac{1}{2}$$

The next step is a clever one: we add some constant to both sides so as to make the left-hand side a perfect square (that is, the left-hand side will be the square of $x + k$, $k$ a constant). With a bit of reflection we come up with the constant to be added, namely "the square of one-half the coefficient of x." In our example this is $(\frac{1}{4})^2$.

*Step 3*    Add $\frac{1}{16}$ to both sides.

$$x^2 + \frac{x}{2} + \frac{1}{16} = \frac{1}{16} + \frac{1}{2}$$

Now another way of writing the left-hand side is shown in step 4.

*Step 4*    Our perfect square is exhibited in the form $(x + \frac{1}{4})^2$.

$$(x + \tfrac{1}{4})^2 = \tfrac{9}{16}$$

This follows since

$$\left(x + \frac{1}{4}\right)^2 = x^2 + \frac{x}{2} + \frac{1}{16}$$

and          $\tfrac{9}{16} = \tfrac{1}{16} + \tfrac{8}{16}$     or     $\tfrac{9}{16} = \tfrac{1}{16} + \tfrac{1}{2}$

*Step 5*    Extract the square root of each side.

$$x + \tfrac{1}{4} = \pm\tfrac{3}{4}$$

Hence          $x = -\tfrac{1}{4} \pm \tfrac{3}{4}$

$$= -1, \tfrac{1}{2} \quad \blacktriangleleft$$

Let us trace these steps in the general case of $ax^2 + bx + c = 0$, because the method applies and we shall be able to solve any quadratic.

*Step 1*    Transpose c.

$$ax^2 + bx = -c$$

*Step 2*    Divide by a.

$$x^2 + \frac{b}{a}x = -\frac{c}{a}$$

*Step 3*    Add to both sides the square of one-half the coefficient of x.

$$x^2 + \frac{b}{a}x + \left(\frac{b}{2a}\right)^2 = \frac{b^2}{4a^2} - \frac{c}{a}$$

*Step 4*    The left-hand side is thus transformed into the square of

$x + b/2a$.  And the right-hand side looks better when reduced to a common denominator.  Thus

$$\frac{b^2}{4a^2} - \frac{c}{a} = \frac{b^2}{4a^2} - \left(\frac{c}{a} \times \frac{4a}{4a}\right)$$

$$= \frac{b^2}{4a^2} - \frac{4ac}{4a^2}$$

$$= \frac{b^2 - 4ac}{4a^2}$$

Therefore
$$\left(x + \frac{b}{2a}\right)^2 = \frac{b^2 - 4ac}{4a^2}$$

(We have reduced the right-hand side to a common denominator.)

*Step 5*   Extract the square root of each side.

$$x + \frac{b}{2a} = \pm \sqrt{\frac{b^2 - 4ac}{4a^2}}$$

$$= -\frac{b}{2a} \pm \frac{1}{2a} \sqrt{b^2 - 4ac}$$

$$= \frac{-b \pm \sqrt{b^2 - 4ac}}{2a}$$

You should memorize this formula.
   We do the arithmetic once more for a numerical example.

◆ **Illustration 2.4**

$$3x^2 - 2x - 6 = 0 \qquad a = 3, b = -2, \text{ and } c = -6 \qquad (2)$$
$$3x^2 - 2x = 6$$
$$x^2 - \tfrac{2}{3}x = \tfrac{6}{3}$$
$$x^2 - \tfrac{2}{3}x + (-\tfrac{1}{3})^2 = \tfrac{1}{9} + 2$$
$$x^2 - \tfrac{2}{3}x + \tfrac{1}{9} = \tfrac{19}{9}$$
$$(x - \tfrac{1}{3})^2 = \tfrac{19}{9}$$
$$x - \tfrac{1}{3} = \pm\tfrac{1}{3}\sqrt{19}$$
$$x = \frac{1 \pm \sqrt{19}}{3} \quad ◆$$

   Now let us apply the formula to the above examples (1) and (2) in Illustrations 2.3 and 2.4, respectively.

► **Illustrations 2.5**

$$2x^2 + x - 1 = 0 \tag{1}$$

$$x = \frac{-b \pm \sqrt{b^2 - 4ac}}{2a}$$

$$= \frac{-1 \pm \sqrt{1 - (4)(2)(-1)}}{2(2)}$$

$$= \frac{-1 \pm 3}{4}$$

$$= -1, \tfrac{1}{2}$$

$$3x^2 - 2x - 6 = 0 \tag{2}$$

$$x = \frac{-b \pm \sqrt{b^2 - 4ac}}{2a}$$

$$= \frac{2 \pm \sqrt{4 - 4(3)(-6)}}{2(3)}$$

$$= \frac{2 \pm 2\sqrt{19}}{6}$$

$$= \frac{1 \pm \sqrt{19}}{3}$$

To check an "answer" we substitute it back into the quadratic equation. We use $(1 + \sqrt{19})/3$.

$$3\left(\frac{1 + \sqrt{19}}{3}\right)^2 - 2\left(\frac{1 + \sqrt{19}}{3}\right) - 6 \overset{?}{=} 0$$

$$3\left(\frac{1 + 2\sqrt{19} + 19}{9}\right) - \frac{2}{3} - \frac{2}{3}\sqrt{19} - 6 \overset{?}{=} 0$$

$$\frac{20 + 2\sqrt{19}}{3} - \frac{2}{3} - \frac{2}{3}\sqrt{19} - \frac{18}{3} \overset{?}{=} 0$$

$$\frac{20}{3} - \frac{2}{3} - \frac{18}{3} + \frac{2\sqrt{19}}{3} - \frac{2\sqrt{19}}{3} = 0 \qquad \text{O.K.} \quad ◄$$

There is a distinct advantage in using the formula. Memorize the formula; you just plug in the numbers and out comes the answer.

The answers to examples (1) and (2) were real numbers but other situations occur. In the first instance, a quadratic might have only one solution.

♦ **Illustration 2.6**

$$x^2 + 2x + 1 = 0 \qquad a = 1, b = 2, c = 1 \tag{3}$$

$$x = \frac{-2 \pm \sqrt{4 - (4)(1)(1)}}{2}$$

$$= \frac{-2 \pm 0}{2} = -1$$

and $-1$ is the only solution. ♦

Second, a quadratic might have no real number as a solution.

♦ **Illustration 2.7**

$$x^2 + x + 1 = 0 \qquad a = b = c = 1 \tag{4}$$

$$x = \frac{-1 \pm \sqrt{1 - (4)(1)(1)}}{2}$$

$$= \frac{-1 \pm \sqrt{-3}}{2}$$

These two numbers do satisfy the equation but they are not real numbers. They belong to the set of numbers called *complex numbers*. ♦

We mention in passing that it might be worth a *little* effort to factor the quadratic equation, since then the solutions would appear immediately. In (1) above it is rather easy to see that the left-hand factors:

$$(x + 1)(2x - 1) = 0$$

Recall that if a product $ab = 0$, then either $a = 0$ or $b = 0$. Therefore it follows that

$$x + 1 = 0, x = -1 \qquad \text{or} \qquad 2x - 1 = 0, x = \tfrac{1}{2}$$

and the solution set is $\{-1, \tfrac{1}{2}\}$.

It is next to impossible to see what the factors are in (2) and in (4). The factored equation in (3) is $(x + 1)(x + 1) = 0$.

Do not spend much time in trying to factor a quadratic equation. General formulas for the solution of the cubic

$$ax^3 + bx^2 + cx + d = 0$$

and the quartic

$$ax^4 + bx^3 + cx^2 + dx + e = 0$$

were derived in the sixteenth century.

In the nineteenth century, the great Norwegian mathematician

Abel proved the astounding theorem that the general equation of degree greater than 4 cannot be solved with a finite number of the usual arithmetic operations. For higher-degree equations we can approximate the solutions to within any reasonable accuracy, especially with the aid of a computer, and this is quite adequate for practical applications. (See Chap. 5.)

**PROBLEMS 2.3**

**1**  Find all numbers which satisfy the equation.

    **a**  $4x - 5 = 3$      **b**  $3x + 5 = 2$      **c**  $7x - 8 = 11$

    **d**  $3x = 1$          **e**  $x - 5 = 2x - 10$    **f**  $\frac{1}{10}x - \frac{3}{5} = 2$

    **g**  $\sqrt{2}x - 3 = 0$   **h**  $\sqrt{2}x + \sqrt{3} = 0$

**2**  Find all numbers which satisfy the equation.

    **a**  $5x + 3 = 13$      **b**  $3x - 1 = 2$          **c**  $22x - 11 = 0$

    **d**  $2x + \sqrt{2} = \sqrt{3}$   **e**  $4x + 1 = 8x + 2$    **f**  $\frac{3}{10}x + \frac{2}{5} = 1$

    **g**  $\sqrt{3}x - 2 = 1$   **h**  $\sqrt{3}x + 5\sqrt{3} = \sqrt{6}x$

**3**  Find all real numbers which satisfy the equation.

    **a**  $x^2 + 4x - 5 = 0$    **b**  $2x^2 + 5x - 3 = 0$   **c**  $2x^2 + 5x + 12 = 0$

    **d**  $6x^2 - x - 1 = 0$     **e**  $(x - 1)(x + 7) = 0$  **f**  $(2x + 5)(x - 3) = 0$

    **g**  $x^2 - 10x + 25 = 0$  **h**  $x^2 + x + 2 = 0$

**4**  Find all real numbers which satisfy the equation.

    **a**  $x^2 + 9x + 14 = 0$     **b**  $x^2 - 9x - 22 = 0$

    **c**  $3x^2 + 16x - 12 = 0$    **d**  $6x^2 + 13x - 5 = 0$

    **e**  $(4x + 3)(3x + 4) = 0$   **f**  $(2x - \sqrt{5})(x) = 0$

    **g**  $3x^2 + 2x + 1 = 0$     **h**  $x^2 - 18x + 81 = 0$

**5**  Write an equivalent equation of the form $(x - a)(x - b) = 0$ for each of Prob. 4a–d.

**6**  How many real numbers satisfy $ax^2 + bx + c = 0$ if $b^2 - 4ac$

    **a**  is greater than 0?
    **b**  is equal to 0?
    **c**  is less than 0?

**2.4**    **INEQUALITIES**

The inequality $2 < 3$ (2 is *less* than 3) is a fact. It is also a fact that $\sqrt{2} > 1$ (the square root of 2 is *greater* than 1). You are familiar with the symbols $<$ (less than) and $>$ (greater than). How about $\leq$ (less than

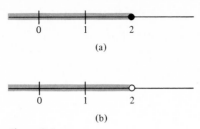

(a)

(b)

**Figure 2.1**

or equal to) and ≥ (greater than or equal to)?  Is 2 ≤ 3 a true statement?
Of course it is.  But $x + 3 \leq 5$ is an *open inequality*.  To solve it we
add $-3$ to both sides, yielding $x \leq 2$.  This is still an open sentence, but
we generally stop here and say that the inequality $x \leq 2$ is satisfied by
the number 2 and any smaller number.  On the number line the set of
numbers satisfying $x \leq 2$ looks like Fig. 2.1a.  It is a half-line with endpoint included.  The graph of the inequality $x < 2$ looks like Fig. 2.1b.
The point $x = 2$ is not included.
    There are rules for deriving equivalent inequalities from a given
one.  These are much like the rules for equations but there is one important difference.  The admissible operations are:

1  If $a < b$, then $a + c < b + c$.
2  If $a < b$ and $c > 0$, then $ac < bc$.
3  If $a < b$ and $c < 0$, then $ac > bc$.

Note that in rule 3 there is a change in the inequality sign.  For example,
$2 < 3$ but $2(-5) > 3(-5)$.  Remember that multiplying both sides of an
inequality by a negative number changes the sign of the inequality.
    The inequality $x - 5 < 0$ is satisfied by each real number less than
5.  An equivalent inequality is $x < 5$.  Sometimes we place two inequality restrictions on a variable.  For example, $x$ might have to be
simultaneously greater than $-3$ and less than 4.  Simultaneously $x$
must be such that $x > -3$ and $x < 4$.  We often write these two inequalities in the form of a double inequality, namely, $-3 < x < 4$.  This
is read, "Negative 3 is less than $x$ and $x$ is less than 4."  Or "$x$ is greater
than negative 3 and less than 4."  On the number line $x$ is any number
greater than $-3$ and less than 4 so the solution set, pictorially, is a line
segment without endpoints (Fig. 2.2).  A segment, or interval, without
its endpoints is called *open*.  One with its endpoints is called *closed*.

**Figure 2.2**

Here are some more linear inequalities with their graphs.

$$2x + 4 \geq 6 \qquad \text{(Fig. 2.3a)}$$
$$x + 8 < 6 \qquad \text{(Fig. 2.3b)}$$
$$3x - 2 \leq 2 \qquad \text{(Fig. 2.3c)}$$
$$x - 2 > -3 \qquad \text{(Fig. 2.3d)}$$

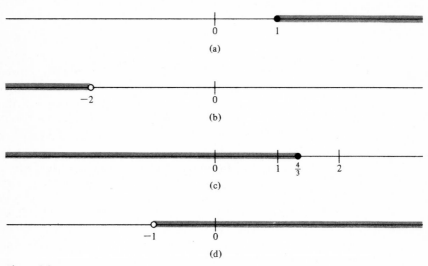

(a)

(b)

(c)

(d)

**Figure 2.3**

Considerable new mathematics concerning systems of linear inequalities have been developed since 1940. The subject matter is known as linear programming and has many applications in business, industry, and economics. We shall touch lightly on it in Sec. 4.8.

Quadratic inequalities are a bit fussier than linear ones. Let us start with a few simple examples.

◆ **Illustration 2.8**

$$x^2 > 0 \qquad \text{(Fig. 2.4a)}$$

This is satisfied by every real number with the exception of 0. ◆

◆ **Illustration 2.9**

$$x^2 \leq 0 \qquad \text{(Fig. 2.4b)}$$

This has the one solution 0. ◆

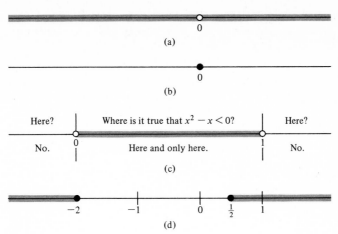

**Figure 2.4**

▶ **Illustration 2.10**

$$x^2 < x \qquad \text{or} \qquad x^2 - x < 0 \qquad \text{(Fig. 2.4c)}$$

In a case like this, it is best first to consider the *equation* $x^2 - x = 0$. The solutions are obtained simply by factoring $x(x - 1) = 0$, so $x = 0$ or 1. The two numbers 0 and 1 divide the number line into three regions: one to the left of the origin, one between 0 and 1, and one to the right of 1.

To the left of 0 (where real numbers are negative) is $x^2 - x < 0$ true? No; $x^2$ is positive and so is $-x$. At $x = 0$, is $x^2 - x < 0$? No. At $x = 1$, is $x^2 - x < 0$? No. Between 0 and 1, is $x^2 - x < 0$? Yes; the square of a number between 0 and 1 is less than the number. To the right of 1, is $x^2 - x < 0$? No; the square of a number larger than 1 is larger than the number so that to the right of 1, $x^2 - x > 0$. So by considering all of these cases we have solved the inequality. The solutions of $x^2 - x < 0$ are all of the numbers between 0 and 1, exclusive of 0 and 1. We write this set as $\{x \mid 0 < x < 1\}$, read "the set of numbers $x$ such that $x$ is greater than 0 and less than 1." We also say that the solution in interval form is $0 < x < 1$ (a line segment without endpoints). ◀

▶ **Illustration 2.11**

$$(x - \tfrac{1}{2})(x + 2) \geq 0 \qquad \text{(Fig. 2.4d)}$$

Again we first consider the equality $(x - \tfrac{1}{2})(x + 2) = 0$ which has $\tfrac{1}{2}$ and $-2$ as solutions. These two numbers divide the number line into three regions: one to the left of $-2$, one between $-2$ and $\tfrac{1}{2}$, and one to the right of $\frac{1}{2}$. We can easily see from the inequality $(x - \frac{1}{2})(x + 2) > 0$ that if $x > \frac{1}{2}$, then each of the two factors is positive and the inequality is therefore true in this region. We can also see that if x lies in between $-2$ and $\frac{1}{2}$, then $(x - \frac{1}{2})$ is negative and $(x + 2)$ is positive. The product of the two factors therefore is negative, and so the inequality is false. For values of x less than $-2$, each factor is negative and their product is positive, and so in this region the inequality is true. The solution set is therefore the union of two sets, namely,

$$\{x \mid x \le -2\} \cup \{x \mid x \ge \tfrac{1}{2}\}$$

We can read this, "The solution set is all numbers less than or equal to $-2$ and all numbers greater than or equal to $\frac{1}{2}$." Geometrically the solution set is two half-lines with endpoints.

## PROBLEMS 2.4

*In Probs. 1 to 4, find the solution set and sketch.*

**1 a** $7x - 21 \le 7$    **b** $6x + 18 \ge -18$    **c** $4x - 3 > 5$
  **d** $2x + 11 < -9$    **e** $3 - 5x \le 6$    **f** $-3x + 4 \ge -1$

**2 a** $9x - 2 \ge 16$    **b** $5x + 5 \le 20$    **c** $x + 4 > 7$
  **d** $2x + 3 < 6$    **e** $7 - 6x \le 8$    **f** $3 - 4x \ge 6$

**3 a** $x^2 - 3x + 2 \le 0$    **b** $x^2 - x - 2 \ge 0$
  **c** $2x^2 + 5x - 3 > 0$    **d** $2 - x - x^2 > 0$

**4 a** $2x^2 - x - 6 \le 0$    **b** $x^2 - 5x + 4 \ge 0$
  **c** $x - x^2 \le 0$    **d** $-x^2 \le 0$

## 2.5    INEQUALITIES WITH ONE ABSOLUTE-VALUE TERM

Inequalities, especially those involving absolute values, pervade mathematics. The basic ideas of calculus are saturated with them and they are found throughout the field of analysis. We shall get only a peek at them.

Recall the definition of $|a|$ given in Sec. 1.9. For real $a$,

$$|a| = \begin{cases} a & \text{if } a \ge 0 \\ -a & \text{if } a < 0 \end{cases}$$

So, with an expression like $|x + b|$, there are two cases to consider:

*Case 1*    If $x + b \ge 0$, then $|x + b| = x + b$.

*Case 2*    If $x + b < 0$, then $|x + b| = -(x + b)$.

Let us now find all numbers which satisfy the inequality

$$|x + 2| - 3 \le 0$$

*Case 1*    If $x + 2 \ge 0$ or, equivalently, if

$$x \ge -2 \tag{1}$$

then $|x + 2| = x + 2$ and the inequality becomes $x + 2 - 3 \le 0$, or

$$x \le 1 \tag{2}$$

Both (1) and (2) must hold simultaneously, and this means that we must take the *intersection* of their solution sets. Thus $\{x \mid x \ge -2\} \cap \{x \mid x \le 1\} = \{x \mid -2 \le x \le 1\}$. The final set can be recorded in interval form as $-2 \le x \le 1$, meaning all numbers in the closed interval from $-2$ to $1$. (See Fig. 2.5a).

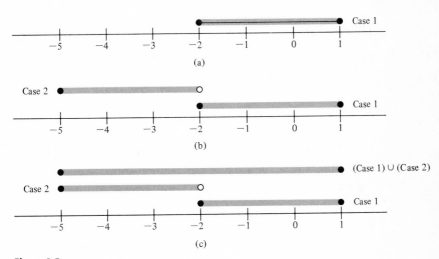

Figure 2.5

*Case 2*    If $x + 2 < 0$, or equivalently, if

$$x < -2 \tag{3}$$

then $|x + 2| = -x - 2$ and the given inequality becomes $-x - 2 - 3 \le 0$, or $-x \le 5$. This can be rewritten in simpler form. We can add $x$ to both sides. This gives $0 \le 5 + x$. Now we add $-5$ to both sides. This gives $-5 \le x$. We can turn this inequality around (reading from right to left), and this gives

$$x \ge -5 \tag{4}$$

All of these operations can be accomplished more readily by multiplying the inequality $-x \leq 5$ by $-1$ and changing the direction of the inequality. ($a < b$ and $-a > -b$ are equivalent.) The intersection of the solution sets of (3) and (4) is (Fig. 2.5b) $\{x|\ x < -2\} \cap \{x\ |\ x \geq -5\} = \{x\ |-5 \leq x < -2\}$. Now any number that is a solution of the given inequality $|x + 2| - 3 \leq 0$ is to be counted whether it arises from case 1 or case 2, and this means that we must take the *union* of the solution sets of the two cases. Thus the solution set of the given inequality is (Fig. 2.5c) $\{x\ |-2 \leq x \leq 1\} \cup \{x\ |-5 \leq x < -2\} = \{x\ |-5 \leq x \leq 1\}$. The resulting solution set can be written in double-inequality form as $-5 \leq x \leq 1$. Any number greater than or equal to $-5$ and less than or equal to 1 satisfies the given inequality (Fig. 2.6).

**Figure 2.6**

If the original inequality had been $|x + 2| - 3 < 0$, the solution would have been $-5 < x < 1$ and the endpoints of the line segment would not have been included (Fig. 2.7).

**Figure 2.7**

▶ **Illustration 2.12**

As another illustration we consider solving $|x + 2| + 3 \leq 0$. Now it is immediately obvious that there is no number satisfying this inequality, because $|x + 2|$ is positive (or 0) so the sum $|x + 2| + 3$ is always positive. But it will be very instructive to set up the two cases as in the previous illustration to see what happens.

*Case 1*   If $x + 2 \geq 0$, or

$$x \geq -2 \tag{5}$$

then $|x + 2| = x + 2$ and the inequality becomes $x + 2 + 3 \leq 0$; or

$$x \leq -5 \tag{6}$$

Since the intersection of (5) and (6) is the null set, there is no solution in this case.

*Case 2*   If $x + 2 < 0$, or

$$x < -2 \tag{7}$$

then $|x + 2| = -(x + 2)$ and the inequality becomes $-x - 2 + 3 \leq 0$, or

$$x \geq 1 \tag{8}$$

Since the intersection of (7) and (8) is the null set; there is no solution in this case either.   So no number satisfies the given inequality.  ◀

**2.6**    **INEQUALITIES WITH TWO ABSOLUTE-VALUE TERMS**

With one absolute-value term $|ax + b|$ present, two cases arise; one where $ax + b \geq 0$ and another where $ax + b < 0$. Now consider an inequality in which there are two absolute-value terms, $|ax + b|$ and $|cx + d|$.  Four cases must be inspected to cover all combinations of signs, as follows:

*Case 1*      $ax + b \geq 0$   and   $cx + d \geq 0$

*Case 2*      $ax + b \geq 0$   and   $cx + d < 0$

*Case 3*      $ax + b < 0$   and   $cx + d \geq 0$

*Case 4*      $ax + b < 0$   and   $cx + d < 0$

However, it will always happen that one of these cases cannot occur: The joint inequality conditions will be inconsistent.  We will see this in the following illustrations.

◀  **Illustration 2.13**

We are to find all numbers that satisfy the inequality

$$|2x| - |x - 2| - 1 \geq 0$$

*Case 1*   If $2x \geq 0$, that is if $x \geq 0$ and if $x - 2 \geq 0$, that is if $x \geq 2$, then the given inequality reduces to $2x - (x - 2) - 1 \geq 0$, or $x \geq -1$.  Now the three inequalities $x \geq 0$, $x \geq 2$, and $x \geq -1$ must hold simultaneously and this means we must take their intersection.  This is

$$\{x \mid x \geq 0\} \cap \{x \mid x \geq 2\} \cap \{x \mid x \geq -1\} = \{x \mid x \geq 2\}$$

(Fig. 2.8a).

*Case 2*   If $2x \geq 0$, that is, if $x \geq 0$ and if $x - 2 < 0$, that is if $x < 2$, then

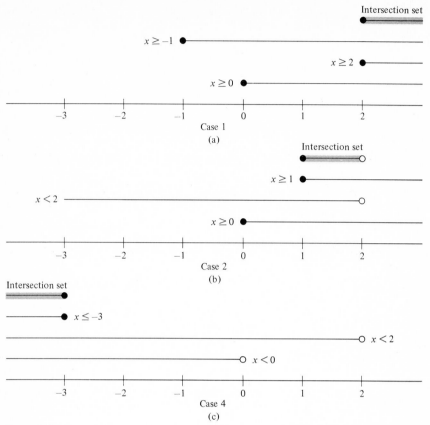

**Figure 2.8**

the given inequality reduces to $2x + x - 2 - 1 \geq 0$, that is to $x - 1 \geq 0$. The intersection of these three sets is the set $\{x \mid 1 \leq x < 2\}$ (Fig. 2.8b).

*Case 3*                $2x < 0$    and    $x - 2 \geq 0$

These inequalities are inconsistent, so this is the impossible case.

*Case 4*   If $2x < 0$, that is if $x < 0$ and if $x - 2 < 0$, that is if $x < 2$, then the given inequality reduces to $-2x + x - 2 - 1 \geq 0$, that is to $-x \geq 3$, or better, $x \leq -3$. The intersection of these three sets is the set $\{x \mid x \leq -3\}$ (Fig. 2.8c).

The union of the sets in cases 1, 2, and 4 is

$$\{x \mid x \geq 2\} \cup \{x \mid 1 \leq x < 2\} \cup \{x \mid x \leq -3\}$$
$$= \{x \mid x \leq -3\} \cup \{x \mid x \geq 1\}$$

Geometrically we have two half-lines with endpoints as in Fig. 2.9 ◀

(Case 1) ∪ (Case 2) ∪ (Case 4)

**Figure 2.9**

▶ **Illustration 2.14**

Let us solve another:

$$|2x - 4| - |x + 3| - 3x + 5 \leq 0$$

*Case 1*        $2x - 4 \geq 0$    and    $x + 3 \geq 0$

Then the inequality reduces to $2x - 4 - x - 3 - 3x + 5 \leq 0$, or $x \geq -1$. The intersection set is $x \geq 2$ (Fig. 2.10a).

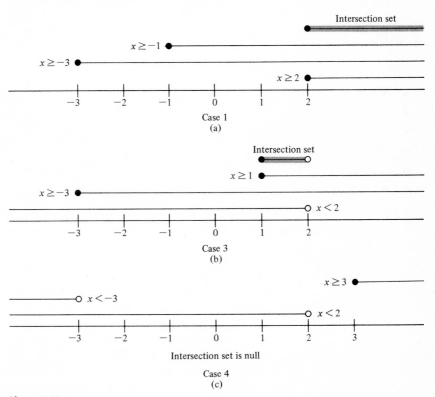

**Figure 2.10**

*Case 2*                    $2x - 4 \geq 0$     and     $x + 3 < 0$

Inconsistent; this case is impossible.

*Case 3*                    $2x - 4 < 0$     and     $x + 3 \geq 0$

Then the inequality reduces to $x \geq 1$. The intersection set is $1 \leq x < 2$ (Fig. 2.10b).

*Case 4*                    $2x - 4 < 0$     and     $x + 3 < 0$

The inequality reduces to $-2x + 4 + x + 3 - 3x + 5 \leq 0$, or $x \geq 3$. The intersection set is the null set.  The union of the sets in the three cases 1, 3, and 4 yields the solution set $\{x \mid x \geq 1\}$ (Fig. 2.11).  ◀

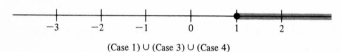

(Case 1) ∪ (Case 3) ∪ (Case 4)

**Figure 2.11**

## PROBLEMS 2.6

*In Probs. 1 to 4, find the solution set and sketch.*

**1** **a** $|x| - 5 \leq 0$           **b** $|x| - 2 \geq 0$           **c** $|x| - 4 \geq 0$
   **d** $|x| - 1 \leq 0$           **e** $|x - 5| - 2 < 0$        **f** $|x + 4| - 3 \leq 0$
   **g** $|2x - 1| + 3 \leq 0$      **h** $|2x - 3| + 1 > 0$

**2** **a** $|x| - 3 > 0$            **b** $|-x| + 3 \leq 0$         **c** $|-2x| + 1 \geq 0$
   **d** $|3x| - 6 < 0$            **e** $|3x + 3| - 6 \leq 0$     **f** $|x + 3| - 4 \leq 0$
   **g** $|5x - 2| - 8 \geq 0$     **h** $|7x + 4| - 11 < 0$

**3** **a** $|5x - 2| - x - 10 \leq 0$     **b** $|2x - 3| + x - 6 \leq 0$
   **c** $|3x + 2| - x + 4 > 0$        **d** $|4x - 1| - x - 2 < 0$
   **e** $|2x| + 2|x| - 8 < 0$         **f** $|x| + |-x| + x - 1 \geq 0$

**4** **a** $|x - 3| - |2x + 3| - 6 \leq 0$   **b** $|x - 1| - |x| - 3 \geq 0$
   **c** $|x - 1| + |x| - 3 < 0$          **d** $|x - 3| - |2x + 3| \geq 0$
   **e** $|2x - 4| - |x + 3| - 3x + 5 \leq 0$   **f** $x^2 - 2|x| + 1 > 0$
                                         Hint: Case 1, $x \geq 0$;
                                               case 2, $x < 0$.

# 3

# FUNCTIONS

## 3.1   FUNCTIONS AND GRAPHS

In Chap. 2 we saw many examples of the graph of a set of numbers pictured as a collection of points on a real number line. Now we begin the central theme of this book—functions. The principal ingredients of the functions we will study are sets of numbers and a relationship between these sets of numbers. This relationship is frequently defined by means of an equation. In this chapter we will study generally what functions are, how to picture (graph) them, and how to combine them. In subsequent chapters we will study specific kinds of functions and their characteristics. Throughout the rest of this book we shall be concerned with the graph of relationships between two interdependent sets. A graph gives us a physical *picture* of a function.

▶ **Illustration 3.1**

Think of the interdependence of automobile fuel consumption and speed. Let us assume that at 10 miles per hour fuel consumption is at the rate of 24 miles per gallon. We write these numbers as an ordered

pair (10, 24), using 10 (miles per hour) as the first member of the ordered pair and 24 (miles per gallon) as the second member.  At 20
miles per hour the consumption is about the same, so the ordered pair
is (20, 24).  As speed increases the number of miles per gallon decreases, say, as in the following ordered pairs: (30, 23), (40, 22), (50, 19),
(60, 16), (70, 13), and (80, 9).  To get a picture, or graph, of this situation we choose two perpendicular lines in a plane, one horizontal and
the other vertical, using the horizontal one for the set of first members
of the ordered pairs and the vertical one for the set of second members.
On the lines, called *axes*, appropriate scales are marked starting with 0
at the intersection of the lines, called the *origin*.  (Of course at 0 miles
per hour the fuel consumption is 0, but we are not here concerned with
speeds of less than 10 miles per hour.)  To plot the first ordered pair
(10, 24) we go horizontally to the right of the origin 10 units and then
vertically 24 units (Fig. 3.1).  Continuing in this manner we locate
eight points in the plane and this is the graph of the assumed data.  If
we imagine that many, many more observations had been made, then
we would have many more points on the graph, which would suggest
that we might sketch a smooth curve through the points.

Let us designate a general ordered pair in this illustration by the
notation $(x, y)$, where $x$ is a variable (speed, in miles per hour) and $y$ is
also a variable (miles per gallon).  Since miles per gallon ($y$) depends
upon miles per hour ($x$), we call $y$ the *dependent variable* and $x$ the
*independent variable*.  More details on graphs are presented in
Chap. 4.  ◀

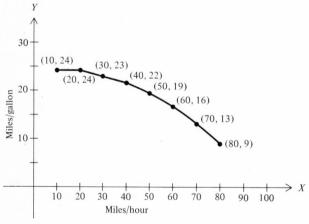

**Figure 3.1**

▶  **Illustration 3.2**

Let $X$ be the set of weights $x$ (in ounces) of letters to be sent by surface

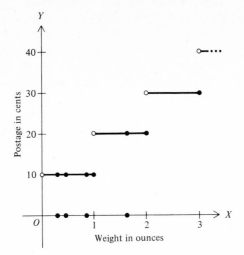

**Figure 3.2**

mail in the Country of Oz and the set Y of postage money y (in cents). For a letter weighing not more than 1 ounce, the postage is 10 cents. For a letter weighing more than 1 ounce but not more than 2 ounces, the postage is 20 cents. The interdependence between elements of set X and those of set Y is further exhibited in Fig. 3.2 and Table 3.1. The letter n stands for a positive integer, x is the independent variable, and y is the dependent variable.

**Table 3.1**

| Weight, x | Postage, y |
|---|---|
| $0 < x \le 1$ | 10 |
| $1 < x \le 2$ | 20 |
| $2 < x \le 3$ | 30 |
| . . . . . . . . . . . . . . . . . . . . . . . . . . | |
| $n < x \le n + 1$ | $10(n + 1)$ |

These two illustrations differ in one somewhat important aspect. In the first each variable is a continuous variable: Both speed and consumption are continuous variables, that is, x, the automobile speed, can be any real number between 0 and some reasonable upper-limit speed (of say 80 miles per hour), and y, the gasoline consumption in miles per gallon, can be any real number between 0 and some reasonable upper limit (of say 24 miles per gallon).

In the second illustration, we have a somewhat different situation. The second illustration has gaps in the dependent variable y. Any weight letter between 0 and 1 ounce (inclusive of 1 ounce) can be mailed for 10 cents. The postage for a letter of 1.001 ounces jumps to 20 cents immediately. The dependent variable $y$ is discontinuous.

These illustrations are similar in a very important aspect. Each is of such a character that whenever a value of the independent variable is specified, a unique value of the dependent variable is determined. To a value of the independent variable there corresponds one and only one value of the dependent variable.

We are now in a position to define a function. We assume two sets $X$ and $Y$, where $x \in X$ (read "$x$ is a member of set $X$") and where $y \in Y$ (read "$y$ is a member of set $Y$").

## DEFINITION 3.1

A function $f$, from $X$ to $Y$, is a rule or correspondence which assigns to each $x \in X$ one and only one $y \in Y$. ◀

This says two things:

**1** To each $x$ there corresponds exactly one $y$.
**2** Some $y$'s may be assigned to more than one $x$ and some $y$'s may not be assigned to any $x$.

The definition of a function can also be worded:

A function is a set of ordered pairs, say $(x, y)$, in which no two of the ordered pairs have the same first element $x$.

As a set of ordered pairs the function $f$ is written $f: (x, y)$, and this is read "the function $f$ whose ordered pairs are $(x, y)$."

We call $y$ the value of the function $f$ corresponding to $x$ and write $y = f(x)$, which is read "$y$ is equal to the value of the function $f$ at $x$" or, more briefly, "$y$ equals $f$ at $x$" or, again, "$y$ equals $f$ of $x$."

Mathematicians distinguish between the function $f$ and the value of the function at $x$, namely $f(x)$. But, unlike some people, they are anxious to get along with as little verbiage as possible and often use $f(x)$ to stand for either the function or its values.

Functions may be determined by tables, graphs, words, or equations. Almost all of our work will deal with equations, usually with $x$ and $y$ as variables, but any letter or symbol will suffice. It is customary to use letters near the end of the alphabet as variables and reserve as constants (parameters) those at the beginning of the alphabet. Often $t$ is used for time, $s$ for speed, $v$ for velocity, etc. A function may be represented by $f$, $F$, $g$, $G$, $h$, $H$, and indeed by the Greek letters $\theta$ (theta), $\phi$ (phi), and $\psi$ (psi). You will no doubt meet with these and perhaps still other notations as you read articles in your own field involving mathematics.

◆ **Illustration 3.3**

Here are some functions defined by equations. Note that we write, indifferently, $y = \cdots$ or $f(x) = \cdots$.

| | |
|---|---|
| $y = ax + b,\ a \neq 0$ | linear function |
| $y = ax^2 + bx + c,\ a \neq 0$ | quadratic function |
| $f(x) = ax^3 + bx^2 + cx + d,\ a \neq 0$ | cubic function |
| $f(x) = ax^n + bx^{n-1} + \cdots + cx + d$ | polynomial function of degree $n$, $n$ a positive integer and $a \neq 0$ |
| $f(x) = \log_{10} x$ | the logarithmic function, $x > 0$ |
| $y = a$ | constant function; for each value of the independent variable, the dependent variable is the constant $a$ |
| $y = \|x\| = \begin{cases} x & \text{if } x \geq 0 \\ -x & \text{if } x < 0 \end{cases}$ | the absolute-value function ◆ |

Here are some functions (and some things that are not functions), with numerical coefficients and with typical computations and remarks.

◆ **Illustration 3.4**

$f(x) = 2x + 3$

$f(1) = 2(1) + 3 = 5$     $f(1)$, read "$f$ at 1," means substitute 1 for $x$ and compute

$f(4) = 2(4) + 3 = 11$     $f(4)$, read "$f$ at 4," means substitute 4 for $x$ and compute

$f(0) = 2(0) + 3 = 3$

$f(-8) = 2(-8) + 3$
$\qquad = -13$

$f(a) = 2a + 3$     $f(a)$ means substitute $a$ for $x$ and compute

$f(a + b) = 2(a + b) + 3$     $f(a + b)$ means substitute $a + b$ for $x$ and compute ◆

▶ **Illustration 3.5**

$$f(x) = 3x^2 - 4x + 5$$
$$f(0) = 3(0)^2 - 4(0) + 5 = 5$$
$$f(2) = 3(2)^2 - 4(2) + 5 = 9$$
$$f(-3) = 3(-3)^2 - 4(-3) + 5 = 44$$
$$f(1 + \sqrt{2}) = 3(1 + \sqrt{2})^2 - 4(1 + \sqrt{2}) + 5$$
$$= 3(1 + 2\sqrt{2} + 2) - 4 - 4\sqrt{2} + 5 = 2\sqrt{2} + 10 \quad ◀$$

▶ **Illustration 3.6**

$$f(x) = \frac{2}{x - 3}$$

$$f(1) = \frac{2}{1 - 3} = -1$$

$$f(2) = \frac{2}{2 - 3} = -2$$

$$f\!\left(\frac{1}{a}\right) = \frac{2}{\dfrac{1}{a} - 3} \qquad a \neq 0, \frac{1}{3}$$

$f(3)$ does not exist since $f(3) = 2/(3 - 3)$.   We never divide by 0.

$$f(-1) = \frac{2}{-1 - 3} = -\frac{1}{2} \quad ◀$$

▶ **Illustration 3.7**

$$f(t) = \sqrt{1 - t^2} \qquad -1 \leq t \leq 1$$
$$f(0) = \sqrt{1 - 0} = 1$$
$$f(\tfrac{1}{2}) = \sqrt{1 - \tfrac{1}{4}} = \tfrac{1}{2}\sqrt{3}$$
$$f(-\tfrac{1}{3}) = \sqrt{1 - \tfrac{1}{9}} = \tfrac{2}{3}\sqrt{2} \quad ◀$$

▶ **Illustration 3.8**

$$f(x) = \log_{10} x \qquad x > 0 \qquad \text{Use Table III, Appendix B}$$
$$f(1) = \log_{10} 1 = 0$$
$$f(2) = \log_{10} 2 = 0.3010$$
$$f(10) = \log_{10} 10 = 1.0000$$
$$f(190) = \log_{10} 190 = 2.2788 \quad ◀$$

▶ **Illustration 3.9**

The indicated function is completely given by the table:

| x    | 0 | 1 | 2  | 4 | 5  |
|------|---|---|----|---|----|
| f(x) | 1 | 4 | -3 | 6 | 19 |

$$f(1) = 4$$
$$f(4) = 6$$
$$f(6) \text{ not defined}$$ ◀

▶ **Illustration 3.10**

The indicated function is completely given by the table:

| f(x) | 0 | 1 | 2  | 4 | 5  |
|------|---|---|----|---|----|
| x    | 1 | 4 | -3 | 6 | 19 |

(Compare with Illustration 3.9.)

$$f(1) = 0$$
$$f(4) = 1$$
$$f(6) = 4$$ ◀

▶ **Illustration 3.11**

The table below does not define a function.

| x    | 1 | 2 | 2 |
|------|---|---|---|
| f(x) | 4 | 4 | 5 |

The same x(=2) yields two different values for $f(x)$. Reread the definition of a function. ◀

▶ **Illustration 3.12**

The equation $y^2 = x$ does not define a function with independent variable x since, for a given x, say x = 4, y has two different values, namely y = 2 and y = −2. But it does define a function with y the independent variable. ◀

## PROBLEMS 3.1

In Probs. 1 to 4, compute $f(1)$, $f(2)$, $f(3)$, and $f(4)$ wherever possible.

**1** $f(x) = \dfrac{2x - 6}{x - 4}$  **2** $f(x) = x - \sqrt{x}$

**3** $f(x) = \dfrac{x + 1}{x^2 + x + 1}$  **4** $f(x) = \sqrt[3]{1 - x}$

In Probs. 5 to 8, compute as indicated.

**5** $s(t) = 16t^2 + 4$.  Find $s(2)$, $s(0)$, $s(-3)$.

**6** $v(\theta) = \dfrac{1}{\theta} - 11$.  Find $v(1)$, $v\left(\dfrac{1}{2}\right)$, $v\left(\dfrac{1}{11}\right)$.

**7** $\phi(x) = \sqrt{1 + x}$, $x \geq 1$.  Find $\phi(1 + \sqrt{2})$, $\phi(3)$.

**8** $f(x) = \dfrac{1}{1 + x}$.  Find $f\left(\dfrac{1}{1 + \sqrt{2}}\right)$.

**9** Does the table below define a function?  If so, what is the independent variable?

| z | 4 | 4 |
|---|---|---|
| $\theta$ | 1 | 2 |

**10** Does the equation below define a function?  If so, what is the independent variable?

$$\phi^2 = t^3$$

**11** Think up an example of a function on your own and describe it verbally.

## 3.2 DOMAIN AND RANGE

It is desirable to examine in more detail the independent variable of a function, especially with regard to its possible values.  Recall that the values form a set.  What are the elements of this set?  Since our work is to involve numbers, the elements will be the set of real numbers or a subset thereof.  This set we call the *domain* of the function.

If the function is defined by the equation $y = 3x + 8$ and there are no special restrictions, then we consider the domain to be the full set of real numbers.

Other situations arise.  If our function is defined by the equation $y = \sqrt{x}$, then automatically $x$ cannot be negative, and the domain is the subset of reals $\{x \mid x \geq 0\}$.  If the equation is $y = 1/(x + 7)$, the domain would usually be taken as $\{x \mid x \neq -7\}$.

Still other cases come to mind.  In quarterly compounding of interest, let $A$ be the amount at the end of the year, $100 the principal, and $r$ the rate of interest (see Sec. 6.1). Then the formula for $A$ is $A = 100(1 + r/4)^4$, and we should consider $r$ the independent variable of the function thus defined. What is the domain? The set of real numbers? Hardly. As a matter of fact a given bank might have three or four different rates, but the domain then is something like, say, $\{0.05, 0.058, 0.065, 0.079\}$.

All of this says that there may be mandatory and/or desirable restrictions on elements of the domain. If so, we so indicate. Otherwise, the domain of a function is to be considered the largest possible subset of the real numbers for which the function is defined.

The values of the dependent variable constitute a subset of the reals called the *range* of the function.

▶ **Illustration 3.13**

Some functions, with domain and range indicated are in Table 3.2.

**Table 3.2**

| Function | Domain | Range |
|---|---|---|
| $f(x) = 2x - 1$ | Reals | Reals |
| $f(x) = 2x^2$ | Reals | Nonnegative reals |
| $f(x) = -\sqrt{x}$ | Nonnegative reals | Nonpositive reals |
| $f(x) = \sqrt[3]{x}$ | Reals | Reals |
| $f(x) = \dfrac{1}{2 - x}$ | Reals except $x = 2$ | Reals except $f(x) = 0$ |

We can determine the range in the last example in the table in the following way. Writing $y$ for $f(x)$ we have $y = 1/(2 - x)$, and we solve this for $x$ in terms of $y$ as in the following steps:

$$y = \frac{1}{2 - x}$$
$$y(2 - x) = 1$$
$$2y - xy = 1$$
$$-xy = 1 - 2y$$
$$xy = 2y - 1$$
$$x = \frac{2y - 1}{y} \qquad \text{and } y \text{ cannot be } 0$$

We see here that there is a value of $x$ for each $y$ except $y = 0$. So each $y$

except 0 is in the range of the function.   It is often difficult to determine the range. ◀

## PROBLEMS 3.2

*In Probs. 1 to 4, determine the domain and range of the defined function.*

**1**  $y = 7x - 20$        **2**  $y = 1 + x^2$

**3**  $f(x) = \sqrt{2x + 3}$      **4**  $y = \dfrac{1}{x}$

*In Probs. 5 to 8, set reasonable limits on the domain and range if the:*

**5**  Horsepower $H$ required for speed, $s$ knots, of a large oil tanker is defined by the equation $H = s^3$.

**6**  Cost $C$ of building a house with $x$ cubic feet is defined by the equation $C = 2x + 800$.

**7**  Number of bacteria $N$ in milk at time $t$, in hours, for the first 24 hours after milking, follows the formula $N = 2^t$.

**8**  Credibility $C$, in percent, of a political figure decreases with the number $n$ of damaging disclosures according to the formula $C = [1/(1 + n^2)]100$.

## 3.3    RULE

In the definition of a function, it was stated that a function from set $X$ to set $Y$ is a rule or correspondence which assigns to each $x \in X$ (domain) one and only one $y \in Y$ (range).   The set of ordered pairs $(x, y)$ could be thought of as the rule, and a function might be defined differently over different sets of real numbers.

### ◆ Illustration 3.14

Consider a function, whose domain is the set of real numbers, defined by the following two equations, each over a different portion of the domain. (See Fig. 3.3.)   There is a single rule, but it is made up of two parts:

$$f(x) = \begin{cases} 0 & x \le 0 \\ 1 & x > 0 \end{cases}$$

To the left of the origin the graph of $y = f(x)$ coincides with the $X$ axis, since $y = 0$ for every $x \le 0$.   To the right of the origin the graph is a line parallel to the $X$ axis and 1 unit above, since $y = 1$ for every $x > 0$. ◀

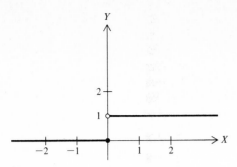

**Figure 3.3**

‣ **Illustration 3.15**

The following is a function, with domain $-1 \leq x \leq 3$, where the defining rule is made up of three parts:

$$f(x) = \begin{cases} x+1 & -1 \leq x \leq 0 \\ 2x & 0 < x \leq 1 \\ 2 & 1 < x \leq 3 \end{cases}$$ ‣

‣ **Illustration 3.16**

Here is an example of a function where the rule is given, verbally, in two parts: (1) The value of the function is 3 diminished by the nearest multiple of 2 if this multiple is odd, and (2) the value of the function is 0 if the nearest multiple of 2 is an even multiple. (A multiple of 2 is written $2k$, where $k$ is an integer, positive, negative, or 0.) Some odd multiples of 2 are $1 \times 2, 3 \times 2, 5 \times 2, -1 \times 2, -11 \times 2$, and so forth. Some even multiples of 2 are $2 \times 2, 4 \times 2, 6 \times 2, 0 \times 2$ (0 is an even integer), $-8 \times 2, -100 \times 2$, and so forth.

We now compute some values of this function. Consider the interval $1 < x < 3$, where the nearest multiple of 2 is $1 \times 2$, and this is an odd multiple since $k = 1$ is an odd number. For every $x$ in this interval the value of the function is $3 - (1)(2) = 1$. Note that the function is not defined at the odd integer points such as $x = \pm 1, \pm 3, \pm 5, \ldots$. For example, take $x = 3$. There is no *nearest* multiple of 2, since both 2 and 4 are multiples of 2. At any odd integer there are *two equally near* multiples of 2. So in the interval $1 < x < 3$ the graph is a line segment, without endpoints, which is parallel to the $X$ axis. In the interval $3 < x < 5$, the nearest multiple of 2 is $2 \times 2$, and this multiple is even and the function is 0. The graph in this interval is a line segment, without endpoints, which coincides with the $X$ axis. In the interval $5 < x < 7$, the nearest multiple of 2 is the odd multiple $3 \times 2$, and the value of the function is $3 - (3)(2) = -3$. The graph is a line segment, without endpoints, parallel to the $X$ axis and 3 units below it. For $-1 < x < 1$ the multiple is even, $0 \times 2$, and the value of the function is 0. For $-3 < x < -1$, the multiple is odd *and* negative, namely $-1 \times 2$, so the value of the function is $3 - (-1)(2) = 5$, and the line segment is 5 units above the $X$ axis. You should now recognize the pattern (Fig. 3.4).

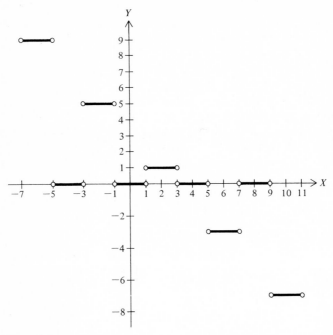

**Figure 3.4**

One might ask, "Can this verbally given function be written with the usual symbols?" The answer is "yes," and if you are curious you can check it out:

$$f(x) = \begin{cases} 3 - 2k & \text{if } (2k - 1) \le x \le (2k + 1),\ k \text{ odd} \\ 0 & \text{if } (2k - 1) \le x \le (2k + 1),\ k \text{ even} \end{cases}$$

**PROBLEMS 3.3**

In Probs. 1 *and* 2, *what is the rule defining the function? What is the independent variable? What is the domain? What is the range?*

| **1** | x | 1 | 2 | 3 |
|---|---|---|---|---|
| | y | 3 | 2 | 1 |

| **2** | v | 0 | 2 | 4 |
|---|---|---|---|---|
| | t | 4 | 2 | 0 |

In Probs. 3 and 4, sketch the function defined by the equations. What is the domain? The range?

**3** $\quad y = \begin{cases} 1 & x < 0 \\ 0 & x = 0 \\ -1 & x > 0 \end{cases}$     **4** $\quad y = \begin{cases} 0 & x < 1 \\ 2 & x > 1 \\ \text{Undefined for } x = 1 \end{cases}$

In Probs. 5 and 6, what is the rule? The domain? The range?

   **5**  See Fig. 3.5.
   **6**  See Fig. 3.6.

In Probs. 7 and 8, what is the rule? The domain? The range?

   **7**  The function is defined as follows: The value of the function at x is the square of the independent variable x diminished by 2 times the cube of it.
   **8**  The function is defined as follows: The value of the function is 2 everywhere except at the integer points k, where it has the value k.

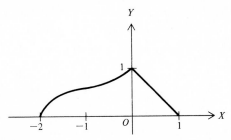

**Figure 3.5**

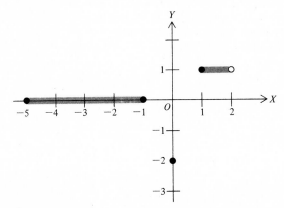

**Figure 3.6**

**3.4    ALGEBRA OF FUNCTIONS**

We have studied the four elementary operations of arithmetic, $+$, $-$, $\times$, and $\div$, in connection with numbers (Chap. 1.). These ideas can also be applied to functions.

Just as the operations of $+$, $-$, $\times$, and $\div$ were applied to pairs of numbers to get a new number, these operations can be applied to pairs of functions to get a new function. The rules for doing this are contained in the following definition.

**DEFINITION 3.2**

Let $f$ and $g$ be functions with domains $d_f$ and $d_g$:

  **a**  $f + g$ is the function defined by $(f + g)(x) = f(x) + g(x)$ for each x in $d_f \cap d_g$
  **b**  $f - g$ is the function defined by $(f - g)(x) = f(x) - g(x)$ for each x in $d_f \cap d_g$
  **c**  $f \times g$ is the function defined by $(f \times g)(x) = f(x) \times g(x)$ for each x in $d_f \cap d_g$
  **d**  $f \div g$ is the function defined by $(f \div g)(x) = f(x) \div g(x)$ for each x in $d_f \cap d_g$ such that $g(x) \neq 0$  ◀

In essence this definition says that you get the values of the sum, difference, product, or quotient of two functions by taking the sum, difference, product, or quotient of their values at points common to their domains.

▶ **Illustration 3.17**

Given the two functions $f$ and $g$, defined by $y = x^3$ and $z = x^2$; the domain of each is the set of real numbers.

Then
$$y + z = x^3 + x^2$$
$$y - z = x^3 - x^2$$
$$yz = x^3 \times x^2 = x^5$$
$$\frac{y}{z} = \frac{x^3}{x^2} = x \qquad x \neq 0$$

[Note that $x = 0$ is not in the domain of $y/z$ since $g(0) = 0$.]  ◀

▶ **Illustration 3.18**

Let $f$ and $g$ be defined by $y = 1 + 1/x$, $z = \sqrt{1 - x^2}$. The domain $d_f$ is the set of all real numbers excluding 0; the domain $d_g$ is the set of all real numbers between $-1$ and $1$, inclusive.

We have:

$$(f + g)(x) = 1 + \frac{1}{x} + \sqrt{1 - x^2} \qquad d_{f+g} \text{ is } -1 \leq x \leq 1, x \neq 0$$

$$(f - g)(x) = 1 + \frac{1}{x} - \sqrt{1 - x^2} \qquad d_{f-g} \text{ is } -1 \leq x \leq 1, x \neq 0$$

$$(fg)(x) = \left(1 + \frac{1}{x}\right)\sqrt{1 - x^2} \qquad d_{fg} \text{ is } -1 \leq x \leq 1, x \neq 0$$

$$\left(\frac{f}{g}\right)(x) = \frac{1 + (1/x)}{\sqrt{1 - x^2}} \qquad d_{f/g} \text{ is } -1 < x < 1, x \neq 0$$

In $f/g$ we must exclude $x = \pm 1$, since $g(-1) = g(1) = 0$. ◀

◆ **Illustration 3.19**

Let $f$ and $g$ be defined by $y = x^2/(1 - x^2)$ and $z = 1/x^2$, respectively, where $d_f$ is reals except $\pm 1$ and $d_g$ is reals except $0$.   Then

$$(f + g)(x) = \frac{x^2}{1 - x^2} + \frac{1}{x^2} \qquad d_{f+g} \text{ is reals except} \pm 1, 0$$

$$(f - g)(x) = \frac{x^2}{1 - x^2} - \frac{1}{x^2} \qquad d_{f-g} \text{ is reals except} \pm 1, 0$$

$$(fg)(x) = \frac{x^2}{1 - x^2} \times \frac{1}{x^2} = \frac{1}{1 - x^2} \qquad d_{fg} \text{ is reals except} \pm 1, 0$$

[Note that the simplification to $1/(1 - x^2)$ is not possible if $x = 0$.]

$$\left(\frac{f}{g}\right)(x) = \frac{x^2}{1 - x^2} \div \frac{1}{x^2} = \frac{x^2}{1 - x^2} \times \frac{x^2}{1} = \frac{x^4}{1 - x^2} \qquad d_{f/g} \text{ is reals except} \pm 1, 0$$

[Note that the simplification to $x^4/(1 - x^2)$ is not possible if $x = 0$.] ◀

## PROBLEMS 3.4

In Probs. 1 to 8, find (a) $d_f$ and $d_g$, (b) $f(x) \pm g(x)$ and $d_{f \pm g}$, (c) $f(x) \cdot g(x)$ and $d_{fg}$, (d) $f(x)/g(x)$ and $d_{f/g}$.

1   $f(x) = 2x - 2$, $g(x) = 3/x$
2   $f(x) = 3x$, $g(x) = 3x$
3   $f(x) = \sqrt{1 - x}$, $g(x) = 3/(x - 4)$
4   $f(x) = \sqrt{1 - x}$, $g(x) = \sqrt{2 + x}$
5   $f(x) = x + 5$, $g(x) = x(x + 5)$
6   $f(x) = 1 + \frac{1}{x}$, $g(x) = 1 - \frac{1}{x}$
7   $f(x) = \sqrt{2 - x}$, $g(x) = \sqrt{-4 + x}$

8  $f(x) = \begin{cases} 0 & x \leq 0 \\ x & x > 0 \end{cases}$     $g(x) = \begin{cases} 1 & x \leq 0 \\ x+1 & x > 0 \end{cases}$

## 3.5    COMPOSITE FUNCTIONS

One further operation in the algebra of functions is of great importance. The idea is best explained by some examples.

♦ **Illustration 3.20**

Suppose a pebble is dropped into a quiet pool of water. A circular ripple of growing radius will result. In appropriate units of time and length, let the radius $r$ be given by $r = 4t$, at time $t$. What is the area $A$ inside the circle at time $t = 5$?

Now the area $A$ is surely a function of $t$, but we at first do not know what the function is. We do know that $A$ is a function of $r$, namely

$$A(r) = \pi r^2$$

but we can compute $A$ from this formula only when we know the radius $r$ when, say, $t = 5$. We *do* know $r = 4t$. Therefore we now compute $r(5) = 4 \times 5$, or $r = 20$. From this we find $A = \pi(20)^2 = 400\pi$ square units. The situation is this: We have two functions $A$ and $r$, where

$$A(r) = \pi r^2$$
$$r(t) = 4t$$

and it should be clear that we can express $A(r)$ directly as a function of $t$ by substituting $r = 4t$ in the formula for area $A$. Thus

$$A = \pi(4t)^2$$
or
$$A = 16\pi t^2$$

and now $A$ is a function of $t$, $A(t)$, and at $t = 5$ we have

$$A(5) = 16\pi(5)^2$$
$$= 400\pi \quad ♦$$

Let us describe this in general. We are given two functions $f$ and $g$. In the above illustration the two functions were $A$ (with independent variable $r$) and $r$ (with independent variable $t$). But now we shift to the usual notations $f$ and $g$, each in terms of $x$. Our interest is first to determine $g$ at some value of $x$, that is, $g(x)$, and then to compute the *value* of $f$ at $g(x)$, that is, $f(g(x))$, read "$f$ at $g$ at $x$." We have no interest in computing $f(x)$ when finding $f(g(x))$.

## DEFINITION 3.3

The composite of $f$ and $g$ is the function $f(g)$ whose values are given by $f(g(x))$. The composite of $g$ and $f$ is the function $g(f)$ whose values are given by $g(f(x))$. ◀

Common notations of the composites of $f$ and $g$ and of $g$ and $f$ are $f \circ g$ and $g \circ f$, respectively. Also we use $d_f$, $d_g$, and $d_{f \circ g}$ for the domains of $f$, $g$, and $f \circ g$, respectively. Similarly the ranges are written $r_f$, $r_g$, and $r_{f \circ g}$.

Note that there is an *ordering* of the two functions: In the first instance we speak of the composite of $f$ and $g$; in the second the order is $g$ and $f$. And usually $f(g)$ and $g(f)$ are not the same. That is, the two composites do not commute.

◀ **Illustration 3.21**

Form $f(g(x))$ and $g(f(x))$, where $f(x) = x^2 - x$ and $g(x) = 3x + 5$.

To get $f(g(x))$ we substitute $3x + 5$ for $x$ in $f(x)$. The result is

$$f(g(x)) = (3x + 5)^2 - (3x + 5)$$
$$= 9x^2 + 27x + 20$$

Now to get $g(f(x))$ we substitute $x^2 - x$ for $x$ in $g(x)$. The result is

$$g(f(x)) = 3(x^2 - x) + 5$$
$$= 3x^2 - 3x + 5 \quad ◀$$

◀ **Illustration 3.22**

Form $f(g(x))$ and $g(f(x))$, where $f(x) = 1/(2 - x)$, $x \neq 2$, and $g(x) = x + 1$.

To get $f(g(x))$ we substitute $x + 1$ for $x$ in $f(x)$. The result is

$$f(g(x)) = \frac{1}{2 - (x + 1)}$$
$$= \frac{1}{1 - x} \quad x \neq 1$$

To get $g(f(x))$ we substitute $1/(2 - x)$ for $x$ in $g(x)$. The result is

$$g(f(x)) = \frac{1}{2 - x} + 1$$
$$= \frac{1 + (2 - x)}{2 - x}$$
$$= \frac{3 - x}{2 - x} \quad x \neq 2 \quad ◀$$

▶ **Illustration 3.23**

Consider the problem of inflating a spherical balloon by pumping in a certain volume of gas per unit of time. Let the radius be given by $r = 1 + \sqrt{t}$. Note that the radius will increase but will increase more slowly as the balloon gets larger. Find the volume of the balloon at time $t$. Again this is a composite function problem. The formula for the volume $V$ of a sphere of radius $r$ is $V = \frac{4}{3}\pi r^3$, and we wish to express $V$ as a function of $t$. The composite $V(r(t))$ is computed as follows:

$$V = \tfrac{4}{3}\pi r^3 = \tfrac{4}{3}\pi(1 + \sqrt{t})^3$$
$$= \tfrac{4}{3}\pi(1 + 3t^{1/2} + 3t + t^{3/2})$$

Check the algebra by computing $(1 + \sqrt{t})(1 + \sqrt{t})$, multiplying the result by $(1 + \sqrt{t})$, and simplifying. ◀

▶ **Illustration 3.24**

Suppose that the blood pressure $p$ of a man under mental stress increases as he is subjected to increasing levels of noise $s$ according to the formula $p = 150 + 3s$. Suppose also that $s$ increases with time $t$ according to the formula $s = 10 + \sqrt{t}$. We disregard units and ask what $p$ will be when $t = 16$. The composite function $p(s(t))$ will yield the answer.

$$p = 150 + 3(10 + \sqrt{t})$$
and at $t = 16$ 
$$p = 150 + 3(10 + 4)$$
$$= 192 \quad ◀$$

In Illustration 3.20 the domain and range of $A$ in terms of $r$ is the nonnegative reals. The domain and range of $r$ in terms of $t$ is also the nonnegative reals and likewise for the composite $A$ in terms of $t$, namely $A(r(t))$. Negative values of $t$, $r$, and $A$ play no role. Similar remarks apply to Illustrations 3.23 and 3.24. In Illustration 3.21 each of $d_f$, $d_g$, $d_{f \circ g}$, $d_{g \circ f}$, and $r_g$ is the set of reals. To determine $r_f$, $r_{f \circ g}$, and $r_{g \circ f}$ is beyond the scope of this text. In Illustration 3.22: $d_f$, reals except 2; $d_g$, reals; $d_{f \circ g}$, reals except 1; $d_{g \circ f}$, reals except 2. Following the method used in Illustration 3.13, we find: $r_f$, reals except 0; $r_g$, reals; $r_{f \circ g}$, reals except 0; $r_{g \circ f}$, reals except 1. To obtain the range $r_{g \circ f}$ we write

$$z = g(f(x)) = \frac{3 - x}{2 - x}$$

and solve for x.  We get

$$x = \frac{2z - 3}{z - 1}$$

and so $z \neq 1$.  The range $r_{g \circ f}$ is therefore the reals except for 1.

In order for $f(g(x))$ to exist, it should be clear that each number x must be such that $g(x)$ is a number in the domain of $f$.  Thus $d_{f \circ g}$ is the set of all x's such that $g(x)$ is contained in the domain of $f$.  Since there may be no such x's, the composite function may not be defined at all.  A simple example is the following:

$$f(x) = \sqrt{x} \qquad\qquad d_f: \text{nonnegative reals}$$
$$g(x) = -1/x^2 \qquad\quad r_g: \text{negative reals}$$
$$f(g(x)) = \sqrt{-1/x^2} \qquad \text{which does not exist}$$

since no value of $g(x)$ is in the domain of $f$.


## PROBLEMS 3.5

*In Probs. 1 to 8, form $f(g(x))$ and state the domain $d_{f \circ g}$.*

1   $f(x) = |x|, g(x) = x^2 - 3x + 1$
2   $f(x) = \sqrt{1 - x^2}, g(x) = 1/x$
3   $f(x) = x^3 - x + 6, g(x) = x^2 + x^{-2}$
4   $f(x) = 1/(1 + x), g(x) = x/(1 + x)$
5   $f(x) = 1/(1 + x), g(x) = 1/(1 + x)$
6   $f(x) = |x|, g(x) = x - |x|$
7   $f(x) = \sqrt{x}, g(x) = x^2$
8   $f(x) = x^2 - 3x + 2, g(x) = k$ (constant)
9   Form $f(f(x))$, where $f(x) = 2x - x^2$.
10   Form $f(f(x))$, where $f(x) = x - 1/x, x \neq 0$.
11   The cost C, in cents, of s feet of fence wire is given by the formula $C(s) = 6s$.  Write a formula for the cost of wire for a three-strand fence around a square field with side m feet.  Your setup should involve the idea of composite functions.
12   Same as Prob. 11 above for a rectangular field m feet long and n feet wide.

# 4

# GRAPHS

## 4.1 GRAPH OF A FUNCTION

Recall that a function from $X$ (the domain) to $Y$ (the range) is a correspondence between the two sets $X$ and $Y$ such that to each $x \in X$ there corresponds one and only one $y \in Y$. To picture this graphically we usually let the set $X$ fall along a horizontal number line (called the $X$ axis) and the set $Y$ fall along a vertical line (called the $Y$ axis). Appropriate scales are chosen and marked off on the axes as in Fig. 4.1. The point marked $O$ is the origin and is usually marked 0 on each scale. If $x = 2$ corresponds to $y = 3$, this is shown by going out to the right 2 units on the $X$ axis and then going up 3 units parallel to the $Y$ axis and placing a point there. The point (call it $A$) is labeled $A(2, 3)$; 2 is called the *abscissa*, 3 is called the *ordinate*, and 2 and 3 together are called the *coordinates* of the point $A$. (Such a system is called a rectangular coordinate system.) The point $A$ lies in the first quadrant (Quadrant I).

Now check point $B(-3, 2)$; it lies in Quadrant II. $C(-4, -3)$ lies in Quadrant III and $D(1, -4)$ in Quadrant IV. The four points $A$,

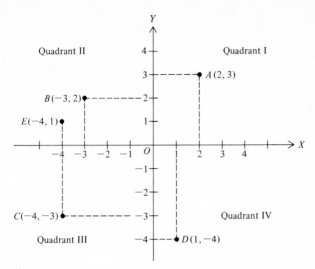

**Figure 4.1**

$B$, $C$, and $D$ constitute a graph and define a simple function which is also defined by Table 4.1, where the correspondence is obvious. If $x$ is considered the independent variable, the domain is the set $X = \{2, -3, -4, 1\}$, and the range is the set $Y = \{3, 2, -3, -4\}$.

**Table 4.1**

| x | 2 | -3 | -4 | 1 |
|---|---|----|----|---|
| y | 3 | 2 | -3 | -4 |

The five points $A$, $B$, $C$, $D$, and $E(-4, 1)$ would not define a function: To $-4$ there would correspond two values of $y$, namely $-3$ and 1, and this is not permitted by the definition of a function.

A function of the type determined by Table 4.1 is called a *discrete* function, being defined for only a few isolated points. But the speed $s$ of an automobile at time $t$ as it accelerates from 0 to 60 miles per hour is a *continuous* function and graphically might look something like Fig. 4.2. There is a technical definition of a continuous function but your intuition must suffice. Loosely speaking a continuous function can be described as a function whose graph can be drawn without lifting the pen. Here for each $t$ such that $0 \le t \le 10$ there corresponds a unique $s$ such that $0 \le s \le 60$.

The graph in Fig. 4.2 was drawn from our knowledge of the characteristics of an automobile and not from a mathematical equation or formula. From the figure we can readily estimate the speed at any selected time $t$ in the interval. For the points indicated we write the ordered pairs $(t, s)$ and incorporate these in Table 4.2. The table does

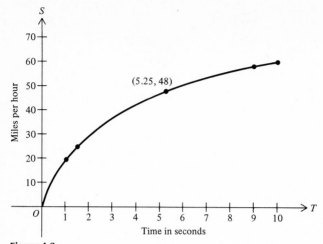

**Figure 4.2**

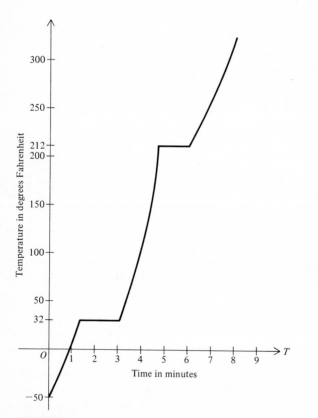

**Figure 4.3**

*not* define the continuous function defined by the graph in Fig. 4.2. The table simply gives a few of the values of *t* and the associated values of *s*.

**Table 4.2**

| *t* | 0 | 1 | 1.5 | 5.25 | 9 | 10 |
|---|---|---|---|---|---|---|
| *s* | 0 | 20 | 25 | 48 | 58 | 60 |

The graph in Fig. 4.3 is that of a continuous function in the domain indicated. It is the kind of graph that arises from applying heat (at a more-or-less constant British thermal units per minute rate) to a block of ice at −50°F. The temperature rises as the ice warms up to 32°F. Then, even though heat is still being applied, the temperature stays constant at 32°F until all of the ice has melted to water. The temperature of the water then rises to 212°F but stops there until all of the water has turned into steam, after which the temperature continues to rise. Note carefully the horizontal portions of the graph. This is the way *water* behaves.

## PROBLEMS 4.1

*In Probs. 1 and 2 state whether the table defines a function and if so what the independent variable is.*

| **1** *t* | 1 | 2 |
|---|---|---|
| *v* | 1 | 2 |

| **2** *w* | 2 | 2 |
|---|---|---|
| *s* | 1 | 2 |

**3** Sketch the triangle whose vertices are $A(-20, 20)$, $B(10, -5)$, and $C(30, 20)$. Find the length of side $AC$.

**4** The graph of some function *f* is shown in Fig. 4.4. From it estimate $f(-3)$, $f(1)$, and $f(\frac{5}{2})$.

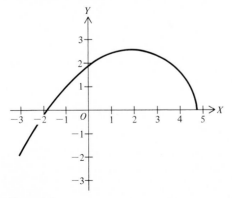

**Figure 4.4**

**5** Given $A(1, 0)$, $B(1, 2)$, and $C(-1, 2)$. If $ABCD$ is a square, find the coordinates of $D$.

**6** Given $A(1, 0)$, $B(1, 2)$, and $C(-1, 2)$. If $ABDC$ is a parallelogram (opposite sides parallel), find the coordinates of $D$.

**7** Given $A(1, 0)$, $B(1, 2)$, and $C(-1, 2)$. If $ADBC$ is a parallelogram (opposite sides parallel), find the coordinates of $D$.

**8** Find the length of the line segment from $A(-3, 6)$ to $B(19, 6)$.

**9** Show that the length of the line segment with one endpoint at $(a, 0)$ and the other endpoint at $(b, 0)$ is $|b - a|$ regardless of what numbers $a$ and $b$ are.

**10** How can you tell by looking at a graph that it is not the graph of a function? In Fig. 4.5 which graphs do not define functions with independent variable x?

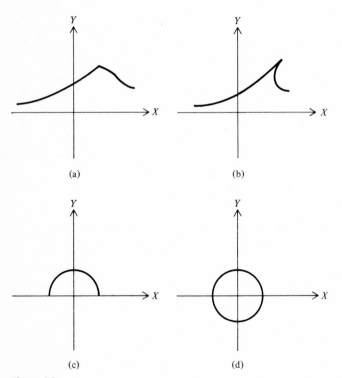

Figure 4.5

## 4.2    GRAPHS OF EQUATIONS AND INEQUALITIES

In this section we plot the graphs of a few equations and inequalities in two variables, one independent and the other dependent. The equations may or may not define functions. Here is a simple example.

▶ **Illustration 4.1**

A graph is to be prepared for $y = 1/x$. Since any number (except 0) can be substituted for $x$, the best we can do is to compute the $y$'s for a small number of $x$'s, preparing first Table 4.3.

**Table 4.3**

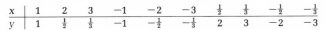

| $x$ | 1 | 2 | 3 | $-1$ | $-2$ | $-3$ | $\frac{1}{2}$ | $\frac{1}{3}$ | $-\frac{1}{2}$ | $-\frac{1}{3}$ |
|---|---|---|---|---|---|---|---|---|---|---|
| $y$ | 1 | $\frac{1}{2}$ | $\frac{1}{3}$ | $-1$ | $-\frac{1}{2}$ | $-\frac{1}{3}$ | 2 | 3 | $-2$ | $-3$ |

Now we plot the ordered pairs $(1, 1)$, $(2, \frac{1}{2})$, etc. as in Fig. 4.6a, which shows the 10 associated points. As yet there is no curve. But if we plot more ordered pairs such as $(\frac{3}{2}, \frac{2}{3})$, $(-2.1, -\frac{10}{21})$, etc., filling in more and more points, we would get something like Fig. 4.6b, and this does suggest a curve. To show that the graph of $y = 1/x$, in Quadrant I is a continuous curve involves some higher mathematics beyond the scope of this text, and similarly for the portion of the graph in Quadrant III. Note carefully that the two branches of the graph are not connected. Such a connection would have to cross the $Y$ axis (where $x = 0$), and $x$ cannot be 0. The domain of the function defined is the set of reals, except 0.

In this case we shall simply connect the isolated points in each branch by drawing something like a smooth curve through them as in Fig. 4.6c. ◀

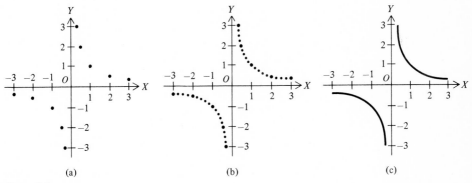

(a)  (b)  (c)

**Figure 4.6**

▶ **Illustration 4.2**

A graph is to be prepared for the equation $y = x + 1$. The computed values of $y$ for selected values of $x$ are shown in Table 4.4. All these ordered pairs (points) are plotted in Fig. 4.7. All appear to lie on a

**Table 4.4**

| x | −2 | −1 | 0 | 1 | 2 | 3 |
|---|----|----|---|---|---|---|
| y | −1 | 0 | 1 | 2 | 3 | 4 |

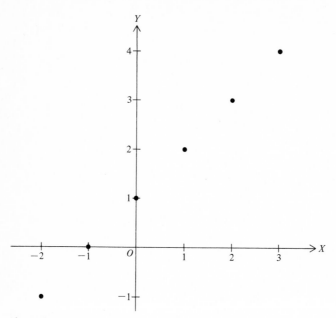

**Figure 4.7**

straight line. They do; and this can be seen as follows (Fig. 4.8a). Let $R(x, y)$ be any point on the line through $P$ and $Q$ (except $P$ and $Q$). Then, from plane geometry, triangles $PQO$ and $PRS$ are similar, and the following proportion holds: $RS/PS = QO/PO$.

Note that the length of $QO$ is 1, and the length of $PO$ is also 1, so that the ratio $QO/PO = \frac{1}{1} = 1$. Also note that $RS = y$ and $PS = x + 1$. Therefore $x$ and $y$ must satisfy the equation $y/(x + 1) = 1$ or $y = x + 1$, and they do since $y = x + 1$ is the equation of the graph. Finally, if $R(x, y)$ is not on the line $PQ$, then $RS/PS \neq QO/PO$ from which it follows that $y \neq x + 1$, which is a contradiction since we are given that $y = x + 1$.

The ideas in Illustration 4.2 may be summarized in the following sentence. A point $(x, y)$ will lie on the graph of $y = x + 1$ if and only if it is a point on the line through the two points $(-1, 0)$ and $(0, 1)$ or any two points in Table 4.4 or any two points computed from the equation $y = x + 1$. But what can we say about the graph of the *inequality* $y < x + 1$? An analysis of this case will provide insight into others.

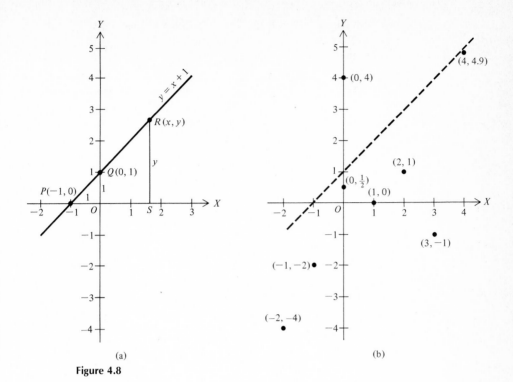

(a)                                    (b)

**Figure 4.8**

♦ **Illustration 4.3**

Sketch the graph of the inequality $y < x + 1$. The best way to see what the graph looks like is to plot the graph of the equality first. This was done in Fig. 4.8a. Now the coordinates of no point on the line $y = x + 1$ will satisfy the inequality. Let us try some points not on the line putting their coordinates into the inequality. Table 4.5 shows an $x$ and a $y$ chosen more or less at random — but not the coordinates of points on the line. It also shows the value of $x + 1$. It also asks the question about these points, "Is $y < x + 1$?." (See Fig. 4.8b, where the graph of $y = x + 1$ is dashed in for reference.) Now it is important to note that each of the seven points chosen for Table 4.5 lie on the

**Table 4.5   Selected points not on the line $y = x + 1$**

| $x$ | −2 | −1 | 0 | 1 | 2 | 3 | 4 |
|---|---|---|---|---|---|---|---|
| $x + 1$ | −1 | 0 | 1 | 2 | 3 | 4 | 5 |
| $y$ | −4 | −2 | $\frac{1}{2}$ | 0 | 1 | −1 | 4.9 |
| Is $y < x + 1$? | Yes | Yes | Yes | Yes | Yes | Yes | Yes |

same side of the line $y = x + 1$ and in each case the inequality $y < x + 1$ is satisfied. Let us test some point on the other side, say $(0, 4)$. The inequality becomes $4 < 0 + 1$, which is false. Moreover the inequality will be false for any point above the line. The general situation suggests itself: The line $y = x + 1$ divides the plane into three regions, one (on the line), where $y = x + 1$, one on one side of the line (called a half-plane), and another on the other side of the line (another half-plane). ◀

**THEOREM 4.1**   *For each point on one side of $y = x + 1$, it is true that $y < x + 1$ and, for each point on the other side, it is true that $y > x + 1$.*

**PROOF**

The proof is immediate. Examine Fig. 4.9a. For any x, $P_1$ is some point below the line, $P_2$ is on the line, and $P_3$ is above the line. The ordinates $y_1 = P_0P_1$, $y_2 = P_0P_2$ and $y_3 = P_0P_3$ are such that the following inequalities obtain:

$$y_1 < x + 1$$
$$y_2 = x + 1$$
$$y_3 > x + 1$$

Check out the details for the points $P_0'$, $P_1'$, $P_2'$, and $P_3'$.

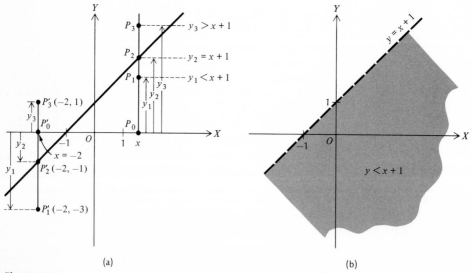

(a)                    (b)

**Figure 4.9**

Therefore the graph of the inequality $y < x + 1$ is the set of points below the line $y = x + 1$ (Fig. 4.9b). The line is not included. It would be included in the graph of $y \leq x + 1$.

These ideas apply to any equation of the form $y = ax + b$, and the graph of such an equation is a straight line, as we shall prove in the next section. In the following illustration we show how the work can be shortened.

► **Illustration 4.4**

Sketch the graph of the inequality $y \geq -2x + 4$. Since $y = -2x + 4$ is the equation of a straight line, and since a straight line is determined by two points, we simply select two points whose coordinates satisfy the equation, plot them, and then draw the line through them. The simplest points would be those on the axes where, for one, $y = 0$, and for the other, $x = 0$. The ordered pairs are $(2, 0)$ and $(0, 4)$, and the line cuts the axes at these two points (Fig. 4.10a). And it can be shown that the above theorem applies equally well to our present equation, so we need only test *one point* to determine the region in which it is true that $y > -2x + 4$. We test at the very simplest point – the origin $(0, 0)$ – since this point is not on the line. We get $0 > -2(0) + 4$ or $0 > 4$, which is false. Therefore on the other side of the line it is true

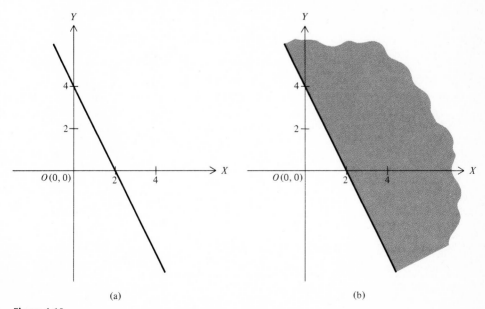

(a)                                              (b)

**Figure 4.10**

that $y > -2x + 4$.  The graph of the inequality $y \geq -2x + 4$ is shaded in Fig. 4.10b, which *includes* the line $y = -2x + 4$.  ◀

## PROBLEMS 4.2

*In Probs. 1 to 6 sketch the graph of the equation in the indicated interval.*

**1**  $y = \dfrac{1}{x^2}, -3 \leq x \leq 3$

**2**  $y = \frac{1}{10}x^3, -3 \leq x \leq 3$

**3**  $y = \dfrac{1}{x - 1}, -2 \leq x \leq 3$

**4**  $y = \dfrac{1}{2 - x}, -1 \leq x \leq 4$

**5**  $x - 2y + 1 = 0, -2 \leq x \leq 2$

**6**  $2x + y - 1 = 0, -1 \leq x \leq 2$

*In Probs. 7 to 12 sketch the graph of the inequality in the indicated interval.*

**7**  $y \leq -x + 2, -3 \leq x \leq 3$

**8**  $y > -2x - 2, -2 \leq x \leq 1$

**9**  $y < x - 3, -1 \leq x \leq 4$

**10**  $4y \geq -x + 2, -2 \leq x \leq 2$

**11**  $x + y + 1 > 0, -2 \leq x \leq 1$

**12**  $x - y - 1 \geq 0, -1 \leq x \leq 2$

## 4.3    LINES

The linear equation in two variables is an equation of the form

$$Ax + By + C = 0 \tag{1}$$

where $A$, $B$, and $C$ are real numbers (not both $A$ and $B$ are 0), and one of the variables, usually $x$, is the independent variable.  This becomes

$$y = -\frac{A}{B}x - \frac{C}{B} \qquad B \neq 0 \tag{2}$$

when solved for $y$ and defines the linear function.
    We wish to prove two statements:

**1**  The graph of $Ax + By + C = 0$ is a straight line.
**2**  Every straight line has an equation of the form $Ax + By + C = 0$.

It is important to consider several special cases in proving statement 1.

*Case 1*    $A = 0$.  Then (2) becomes $y = -C/B$, a constant, and the graph looks like Fig. 4.11.  It is the graph of a constant function—a line parallel to the $X$ axis.

*Case 2*    $B = 0$ [in equation (1), not in (2)].  Then (1) becomes $x = -C/A$ for any value of $y$.  Therefore the graph is a line parallel to the $Y$ axis and is not the graph of a function (Fig. 4.12).

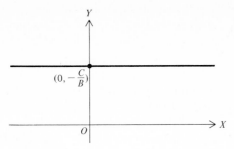

Figure 4.11

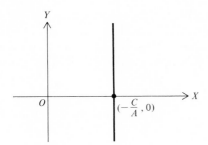

Figure 4.12

For the next two cases we first simplify equation (2), writing $a = -A/B$ and $b = -C/B$. Thus

$$y = ax + b \qquad (3)$$

This is possible since $-A/B$ and $-C/B$ are just two real numbers.

*Case 3* $b = 0$. Then (3) becomes $y = ax$. Now follow the next argument by examining Fig. 4.13. The "curve" of $y = ax$ passes through the origin $(0, 0)$. The equation can be written in the form $y/x = a$,

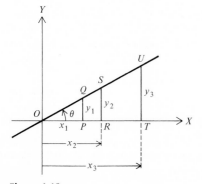

Figure 4.13

$x \neq 0$, and this says that for any sequence of $x$'s, say $x_1$, $x_2$, $x_3$, . . . , and their associated $y$'s, namely $y_1$, $y_2$, $y_3$, . . . , it is true that

$$\frac{y_1}{x_1} = \frac{y_2}{x_2} = \frac{y_3}{x_3} = \cdots = a \qquad \text{a constant}$$

Thus, from plane geometry, triangles $OPQ$, $ORS$, $OTU$, . . . are similar. Therefore $O$, $Q$, $S$, $U$, . . . lie on the straight line whose equation is $y = ax$. The constant $a$ is called the *slope* of the line. Trigonometrically the constant $a$ is the tangent of the angle $\theta$. We take this up in Chap. 9.

*Case 4*   The graph of the equation $y = ax + b$ is a line parallel to the graph of $y = ax$, because for any given abscissa $x$, we get the ordinate to $y = ax + b$ by adding the constant $b$ to the ordinate to $y = ax$ (Fig. 4.14). (Note that the slope of $y = ax + b$ is the constant $a$. When any linear equation is solved for $y$, the coefficient of $x$ is the slope.)

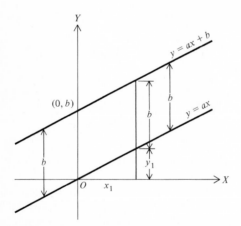

**Figure 4.14**

To prove (statement 2) we refer to Fig. 4.15.   Let $P(x_1, y_1)$ and $Q(x_2, y_2)$ be two distinct fixed points on the given line.   Any two distinct points such as $P$ and $Q$ determine a unique line.   By similarity of triangles the point $R(x, y)$ will lie on the line $PQ$ if and only if the following proportion holds:

$$\frac{y - y_1}{x - x_1} = \frac{y_2 - y_1}{x_2 - x_1} \tag{4}$$

That is, (4) must hold for every real number $x$ ($\neq x_1$) and the associated number $y$.   This means that (4) is an equation of the straight line, and it is called the *two-point form*.

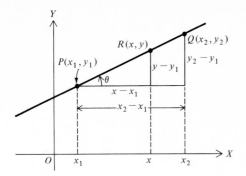

**Figure 4.15**

Equation (4) reduces in the following way.

$$(x_2 - x_1)(y - y_1) = (y_2 - y_1)(x - x_1)$$
$$(y_1 - y_2)x + (x_2 - x_1)y + (x_1y_2 - x_2y_1) = 0$$

This is of the form $Ax + By + C = 0$, so that statement 2 is proved.

Equation (4) is the equation of the line determined by any two points $P(x_1, y_1)$ and $Q(x_2, y_2)$. Reexamine Fig. 4.15 and equation (4). The quantity $(y_2 - y_1)/(x_2 - x_1)$, usually called $m$, is the trigonometric tangent of the angle $\theta$ and, as we have seen, is called the slope of the line.

Thus

$$\frac{y_2 - y_1}{x_2 - x_1} = m = \tan \theta$$

It measures the steepness of the line (see Sec. 9.4).

Reexamine equation (4). A unique line with this equation is determined by the two given points $(x_1, y_1)$ and $(x_2, y_2)$. Since $(y_2 - y_1)/(x_2 - x_1)$ is just one number, namely $m$, equation (4) determines a unique line when one point, say $(x_1, y_1)$, and the slope $m$ are given. In this case we write (4) in the point-slope form as follows:

$$\frac{y - y_1}{x - x_1} = m$$

or
$$y - y_1 = m(x - x_1) \tag{5}$$

▶ **Illustration 4.5**

The equation of the line passing through $(-1, 5)$, $(0, 8)$ is $(y - 5)/(x + 1) = (8 - 5)/(0 + 1)$ or $3x - y + 8 = 0$. The line perpendicular to the $X$ axis and passing through $(4, 17)$ is $x = 4$. The line through the origin with slope $-2$ is $y = -2x$.  The line through $(0, 6)$ with slope $-2$ is $y - 6 = -2(x - 0)$ or $y = -2x + 6$.  The line through $(0, -8)$ with slope $-2$ is $y = -2x - 8$.  The line through $(0, 0)$ with slope $0$ is $y = 0$ (the $X$ axis).  The line through $(-8, 16)$ and $(2, 16)$ is $y = 16$.  ◀

## PROBLEMS 4.3

*In Probs. 1 to 6 find the equation of the line through the indicated points.*

**1** $(1, 2)$ and $(4, -6)$     **2** $(3, -4), (-7, 2)$     **3** $(0, 6), (4, 6)$

**4** $(8, 0), (-4, 0)$     **5** $(3, 3), (2, -7)$     **6** $(2, 0), (2, -19)$

*In Probs. 7 to 10 find the equations of the indicated lines.*

**7** Parallel to the $Y$ axis and passing through $(6, 72)$.

**8** Perpendicular to the $X$ axis and passing through $(-6, 6)$.

**9** Perpendicular to the $Y$ axis and passing through $(-2, -3)$.

**10** Parallel to the $X$ axis and passing through $(5, 3)$.

*In Probs. 11 to 14 find the equation of the indicated line.*

**11** Passing through $(5, -3)$ with slope 2.

**12** Passing through $(0, 0)$ with slope $-3$.

**13** Passing through $(1, 3)$ with slope $\frac{5}{3}$.

**14** Passing through $(0, 4)$ and $(5, 0)$.

## 4.4 SIMULTANEOUS LINEAR EQUATIONS AND INEQUALITIES

Two nonparallel lines intersect at a unique point.  To find the coordinates of this point we seek one ordered pair of numbers $(x_0, y_0)$ which satisfies simultaneously the equation of each line.  The process is called *solving simultaneous equations*.  As an example consider the following.

▶ **Illustration 4.6**

Solve simultaneously:

$$2x + 5y - 3 = 0 \qquad\qquad (1)$$
$$x - y - 2 = 0 \qquad\qquad (2)$$

To solve these first multiply equation (2) by 2 in order to match the coefficient of $x$ in equation (1).

$$2x + 5y - 3 = 0 \qquad (1)$$
$$2x - 2y - 4 = 0 \qquad (2')$$

Subtract (2') from (1).

$$7y + 1 = 0$$

or
$$y = -\tfrac{1}{7}$$

From (2), substituting for y,

$$x + \tfrac{1}{7} - 2 = 0$$

or
$$x = \tfrac{13}{7}$$

The common solution to the pair of equations is $(\tfrac{13}{7}, -\tfrac{1}{7})$. The two lines are sketched in Fig. 4.16. ◀

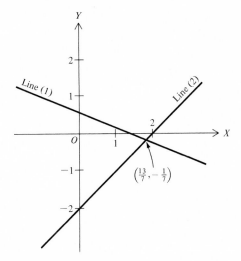

**Figure 4.16**

◆ **Illustration 4.7**

Solve simultaneously:

$$x + 3y - 2 = 0 \qquad (3)$$
$$4x + 12y + 1 = 0 \qquad (4)$$

We follow the method of Illustration 4.6, multiplying equation (3) by 4 in order to make the coefficients of x the same.

$$4x + 12y - 8 = 0 \qquad (3')$$
$$4x + 12y + 1 = 0 \qquad (4)$$

Subtracting (4) from (3′), we get into trouble, because both the x and the y terms disappear, leaving a false statement, namely

$$-9 = 0$$

If we had checked the slopes of the two lines, we would have found that the slopes are the same, each being $-\frac{1}{3}$. The lines are therefore parallel, and parallel lines do not intersect, so the two given equations do not have a common solution (Fig. 4.17). ◀

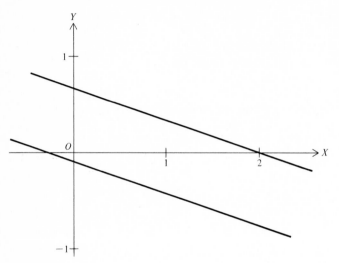

**Figure 4.17**

▶ **Illustration 4.8**

Solve simultaneously:

$$x + 7y - 8 = 0 \qquad (5)$$
$$3x + 21y - 24 = 0 \qquad (6)$$

If we multiply (5) by 3, we get equation (6). The two given equations *are* different but they are equivalent. Each is an equation of the same straight line and any solution of one is a solution of the other. In this case we do not solve them simultaneously. We take one equation, say (5), let x or y be any real number, and then solve for the other variable. In this case it is a little simpler to let y be any real number. Then $x = 8 - 7y$, so that each point on (5) has an ordered pair of the form $(8 - 7y, y)$, where y is any real number. Although there is but one line involved (some authors like to think of the problem as involving two coincident lines), we can still think of the simultaneous solution as being the set of all ordered pairs of the form $(8 - 7y, y)$. ◀

It is easy to extend the ideas of linear inequalities in Sec. 4.2 to simultaneous linear inequalities.  Here are some examples.

### ▶ Illustration 4.9

Solve the simultaneous linear inequalities

$$2x + 5y - 3 \le 0 \tag{7}$$
$$x - y - 2 > 0 \tag{8}$$

We first plot the lines given by the equations.  Line (8) is dashed, since the equality mark is not included (Fig. 4.18).  These lines divide the plane into four regions.  (We are now not counting the lines themselves as regions.)  We test inequality (7) by substituting the coordinates of one point — any point not on the line — in order to determine on which side of line (7) the inequality holds, and then repeat the process for line (8).  Since neither line passes through the origin, we use the point (0, 0) for both.

$$2(0) + 5(0) - 3 \le 0? \tag{9}$$
$$0 - 0 - 2 > 0? \tag{10}$$

In (9) the inequality is true, and in (10) it is false.  In Fig. 4.18 we shade the proper sides of the two lines, and all points in the darkest region constitute the solution of the simultaneous inequalities.  ◀

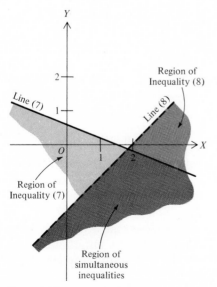

**Figure 4.18**

We have seen in Illustration 4.7 that parallel lines do not intersect, and so the equations do not have a common solution. But suppose we change the equations to inequalities, as in the next illustration. Would there necessarily be a solution? The answer is "no." Whereas two intersecting lines divide the plane into four regions, two parallel lines divide the plane into only three regions. But for two inequalities there are four possible combinations. (For the moment it is not important to include the equality symbol.)

First inequality:    $> 0$    $> 0$    $< 0$    $< 0$
Second inequality:    $> 0$    $< 0$    $> 0$    $< 0$

Therefore one of the inequality combinations is impossible in the case of parallel lines. We have seen a similar situation before in Sec. 2.6.

▶ **Illustration 4.10**

Solve the simultaneous linear inequalities

$$2x - y - 1 \geq 0 \tag{11}$$
$$4x - 2y - 7 \geq 0 \tag{12}$$

The equations are those of parallel lines since they have the same slope, namely 2. Testing at $(0, 0)$ we find $-1 \geq 0$ and $-7 \geq 0$, and both are false. This means we must go to the other side of both lines, and so the solution is the set of all points on or below line (12). The solution set is darker in Fig. 4.19. The solution to the system

$$2x - y - 1 \geq 0 \tag{13}$$
$$4x - 2y - 7 \leq 0 \tag{14}$$

is the set of points between the two lines (Fig. 4.20).

But for the system

$$2x - y - 1 \leq 0 \tag{15}$$
$$4x - 2y - 7 \geq 0 \tag{16}$$

the solution set for the single inequality (15) and the solution set for the single inequality (16) are nonoverlapping. Hence the simultaneous inequalities have no solution (Fig. 4.21). ◀

In general *three* lines do not intersect at a point, and so we do not solve a system of three linear equations simultaneously. But a system of three simultaneous inequalities might very well have a solution. The following is an example.

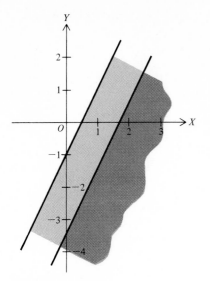

**Figure 4.19**

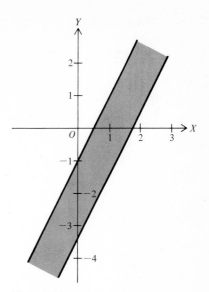

**Figure 4.20**

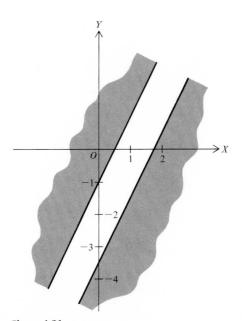

**Figure 4.21**

♦ **Illustration 4.11**

Solve the system of inequalities and sketch.

$$x - y + 1 \geq 0 \qquad (17)$$
$$x + y - 2 \geq 0 \qquad (18)$$
$$3x - y - 3 \leq 0 \qquad (19)$$

Testing each of these inequalities at (0, 0) yields

$$0 - 0 + 1 \geq 0 \qquad \text{true} \qquad (17')$$
$$0 + 0 - 2 \geq 0 \qquad \text{false} \qquad (18')$$
$$0 - 0 - 3 \leq 0 \qquad \text{true} \qquad (19')$$

The shaded set of points constitutes the simultaneous solution set (Fig. 4.22).

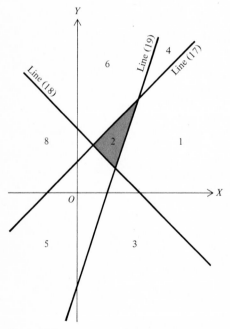

**Figure 4.22**

It is not hard to test this system with all possible inequality combinations. These are eight in number; think of (17), (18), and (19) and the eight possible problems 1 to 8 as on page 81.

| 1 | 2 | 3 | 4 | 5 | 6 | 7 | 8 |
|---|---|---|---|---|---|---|---|
| $\geq 0$ | $\geq 0$ | $\geq 0$ | $\leq 0$ | $\geq 0$ | $\leq 0$ | $\leq 0$ | $\leq 0$ |
| $\geq 0$ | $\geq 0$ | $\leq 0$ | $\geq 0$ | $\leq 0$ | $\geq 0$ | $\leq 0$ | $\leq 0$ |
| $\geq 0$ | $\leq 0$ | $\geq 0$ | $\geq 0$ | $\leq 0$ | $\leq 0$ | $\geq 0$ | $\leq 0$ |

The solution set to any one of these corresponds to one of the regions of the plane cut off by the three lines—well, almost. Three lines divide the plane up into at most seven regions, so one sign set will not have a solution set, column 7 in this case. If the three lines enclose a triangle, the missing region has sign sets opposite those in the triangular region. In the above example column 2 had the sign set $\geq 0$, $\geq 0$, $\leq 0$, and the missing region corresponds to 7 with sign set $\leq 0$, $\leq 0$, $\geq 0$. Figure 4.22 shows the regions corresponding to the different sign sets. ◀

## PROBLEMS 4.4

*In Probs. 1 to 20 draw lines (appropriately solid or dashed) and shade the solution set.*

**1** $x - y + 2 < 0$     **2** $x - y + 2 \geq 0$     **3** $y - 1 < 0$

**4** $y \geq 1$     **5** $x + 3 \geq 0$     **6** $3x + 3 > 0$

**7** $x - y + 4 \leq 0$     **8** $x - y + 4 \leq 0$     **9** $x + y - 5 > 0$
    $x + y - 2 \geq 0$         $x + y - 2 \leq 0$         $2x + y + 2 \geq 0$

**10**   $3x - y \geq 0$     **11** $x - 2y - 7 \leq 0$     **12** $x - 2y - 7 > 0$
    $x + y - 6 \geq 0$         $2x - y + 1 \geq 0$         $2x - y + 1 < 0$

**13**   $x + y - 5 \leq 0$     **14**   $3x + y - 8 > 0$     **15**   $x + y - 5 < 0$
    $2x + y + 2 \leq 0$         $x - 2y + 2 \leq 0$         $5x - y - 5 \leq 0$
        $y - 3 \geq 0$         $2x + 3y - 24 \leq 0$         $x + 8 \geq 0$

**16** $x - y + 4 > 0$     **17** $2x + y - 3 > 0$     **18**   $x + y - 5 < 0$
    $x + y - 2 > 0$         $x - 2y + 1 < 0$         $2x + y + 2 < 0$
        $x - 2 \leq 0$         $x + y - 5 < 0$         $y - 1 < 0$

**19** $x + y + 1 \geq 0$     **20** $x - y \geq 0$
        $x + y \leq 0$         $x - 1 \geq 0$
    $x + y - 1 \leq 0$         $x - 2 \leq 0$

## 4.5    THE DISTANCE FORMULA IN THE PLANE

Points $P(3, 0)$ and $Q(7, 0)$ lie on the $X$ axis, and the distance $PQ$ is simply $7 - 3$, or 4. Points $R(-8, 0)$ and $S(-3, 0)$ lie on the $X$ axis, and the distance $RS$ is $-3 - (-8)$, or 5. Check this. To get $RS$ you subtract the smaller $(-8)$ from the larger $(-3)$.

But for $P(a, 0)$ and $Q(b, 0)$ we do not know whether $a$ or $b$ is the larger. Therefore to compute the distance $PQ$ we subtract in either order and take the absolute value. Thus the distance $PQ$ is $|a - b|$ or $|b - a|$, whichever way we wish to write it.

The distance $PQ$, where the coordinates are

$P(0, 8)$, $Q(0, 15)$ is $15 - 8$, or 7

$P(0, -11)$, $Q(0, -20)$ is $-11 - (-20) = 9$

$P(0, c)$, $Q(0, d)$ is $|d - c|$

$P(4, c)$, $Q(4, d)$ is $|d - c|$

$P(a, c)$, $Q(b, c)$ is $|a - b|$

The above treatment applies only to points on a line parallel to or coincident with one of the axes. In order to find the distance for any two points in the plane we must make use of the pythagorean theorem: The square of the hypotenuse of a right triangle is equal to the sum of the squares of the other two sides (Fig. 4.23).

Now consider the two points $P(x_1, y_1)$ and $Q(x_2, y_2)$. In Fig. 4.24 in the right triangle $PQR$, the length of $PR$ is $|x_2 - x_1|$, and the length of $QR$ is $|y_2 - y_1|$. (In the figure the points are positioned so as to indicate that $x_2 > x_1$ and $y_2 > y_1$, but the formula below for distance holds regardless of the location of the points.)

The distance $PQ$ or $d$ can now be computed from the pythagorean theorem:

$$d^2 = |x_2 - x_1|^2 + |y_2 - y_1|^2 \tag{1}$$

since we know the lengths of two sides. The absolute-value signs can be dropped, since $(x_2 - x_1)^2$ and $(y_2 - y_1)^2$ are positive anyway. Hence the distance $d$ is given by

$$d = \sqrt{(x_2 - x_1)^2 + (y_2 - y_1)^2}$$

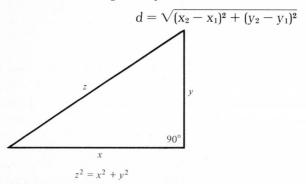

$$z^2 = x^2 + y^2$$

**Figure 4.23**

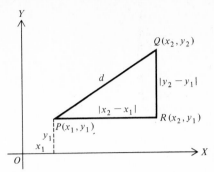

**Figure 4.24**

If, in triangle PQR, R is a right angle, then the pythagorean condition

$$\overline{PQ}^2 = \overline{PR}^2 + \overline{QR}^2 \tag{2}$$

holds. This we have just used above. And, conversely, if condition (2) holds, then angle PRQ is a right angle. Use this converse to work Probs. 9 and 10 of Sec. 4.5.

## PROBLEMS 4.5

*In Probs. 1 to 8, find the distance between the two given points.*

1  (3, 2), (−5, 7)     2  (2, −3), (4, 2)     3  (1, −8), (8, −1)

4  (6, 2), (0, 1)     5  (0, 3), (−3, 0)     6  (8, 2), (12, 2)

7  (4, 9), (4, −9)     8  (5, 2), (−2, −7)

9  Show that the triangle A(0, 0), B(2, 1), C(1, 3) is a right triangle.

10  Show that the triangle A(−3, 0), B(4, −7), C(5, 8) is not a right triangle.

11  Show that the diagonals of the rectangle A(0, 0), B(a, 0), C(a, b), D(0, b) are equal.

12  Show that A(5, −1) is the midpoint of the line segment B(4, −4), C(6, 2).

## 4.6    CIRCLES

Figure 4.25 indicates how the equation of the circle with center at (0, 0) and radius r can be written down immediately by the use of the pythagorean theorem. Let P(x, y) be any point on the circle and C(0, 0) the center. Then the distance r(=PC) is given by

$$r = \sqrt{(x - 0)^2 + (y - 0)^2}$$

or                    $$x^2 + y^2 = r^2 \tag{1}$$

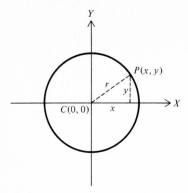

**Figure 4.25**

which must hold for all points on the circle and for no others. So (1) is the equation of the circle. Note that both x and y are raised to the second power. The equation, in each variable, is of the second degree.

And it is just about as easy to write down the equation of the circle with center at (h, k) and radius r (Fig. 4.26). Using the pythagorean theorem on right triangle PCQ we get

$$(x - h)^2 + (y - k)^2 = r^2 \tag{2}$$

and this is the equation of the circle. Again it is of the second degree in x and y.

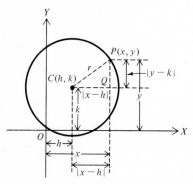

**Figure 4.26**

### ◗ Illustration 4.12

Here are some numerical examples.

$$(x - 2)^2 + (y - 3)^2 = 16 \tag{1}$$

is the equation of the circle with center at (2, 3) and radius 4.

$$(x - 3)^2 + (y + 8)^2 = 25 \tag{2}$$

is the equation of the circle with center at $(3, -8)$ and radius 5.

$$x^2 + (y + 6)^2 = 100 \tag{3}$$

has center $(0, -6)$, radius 10.

$$(x + 6)^2 + (y + 19)^2 = 39 \tag{4}$$

has center $(-6, -19)$, radius $\sqrt{39}$.

Any one of these equations can be expanded. Thus (1) becomes

$$x^2 - 4x + 4 + y^2 - 6y + 9 = 16$$

which simplifies to

$$x^2 - 4x + y^2 - 6y - 3 = 0 \tag{1'}$$

Conversely, equation (1') can be written in form (1). It is desirable to do this if form (1') is given in the first place, and the reason is that form (1) shows the center and radius and form (1') does not. To transform (1') into (1) we complete the square first on $x^2 - 4x$ and then on $y^2 - 6y$. (Review Sec. 2.3.) To $x^2 - 4x$ we add the square of half the coefficient of $x$, getting $x^2 - 4x + 4$. Next we add the square of half the coefficient of $y$, getting $y^2 - 6y + 9$. Now the left-hand side of (1') reads

$$x^2 - 4x + 4 + y^2 - 6y + 9 - 3$$

and we have added a total of 13 to the left-hand side, so we must add 13 to the right-hand side to maintain equality. The equation now reads

$$(x^2 - 4x + 4) + (y^2 - 6y + 9) - 3 = 13$$

or

$$(x - 2)^2 + (y - 3)^2 = 16$$

on transposing the 3 and factoring. This recovers equation (1). ◀

◀ **Illustration 4.13**

Try this one: Find the center and radius of the circle whose equation is $x^2 - 3x + y^2 + 5y - 8 = 0$. Complete the squares:

$$x^2 - 3x + (\tfrac{3}{2})^2 + y^2 + 5y + (\tfrac{5}{2})^2 - 8 = \tfrac{9}{4} + \tfrac{25}{4}$$

$$(x^2 - 3x + \tfrac{9}{4}) + (y^2 + 5y + \tfrac{25}{4}) = 8 + \tfrac{9}{4} + \tfrac{25}{4}$$

$$= \tfrac{66}{4}$$

$$= \tfrac{33}{2}$$

or

$$(x - \tfrac{3}{2})^2 + (y + \tfrac{5}{2})^2 = \tfrac{33}{2}$$

(See Fig. 4.27.) ◀

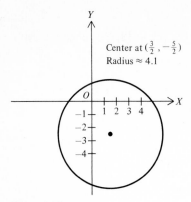

**Figure 4.27**

## PROBLEMS 4.6

*In Probs. 1 to 8 find the equation of the circle.*

**1**  $C(2, 5)$, $r = 6$.          **2**  $C(-4, 8)$, $r = 2$.

**3**  $C(0, -8)$, $r = 1$.          **4**  $C(5, 0)$, $r = 5$.

**5**  $C(2, 2)$ and tangent to the axes.

**6**  $C(-5, 6)$ and tangent to the $X$ axis.

**7**  $C(1, 1)$ and passing through $(0, 0)$.

**8**  $C(2, 2)$ and passing through $(1, 1)$.

*In Probs. 9 to 14 sketch the circle.*

**9**  $x^2 + y^2 - 9 = 0$          **10**  $x^2 + y^2 - 64 = 0$

**11**  $(x - 3)^2 + (y + 4)^2 = 1$          **12**  $(x + 2)^2 + (y - 5)^2 = 4$

**13**  $x^2 - 2x + y^2 + 4y + 1 = 0$.   Hint: Complete the squares to find the center and radius.

**14**  $x^2 + 6x + y^2 - 8y + 16 = 0$.   Hint: Complete the squares to find the center and radius.

*In Probs. 15 to 18 sketch the circles and shade the solution sets.*

**15**  $x^2 + y^2 - 49 \le 0$          **16**  $x^2 + y^2 - 64 \ge 0$

**17**  $x^2 + (y - 2)^2 > 4$          **18**  $(x + 1)^2 + y^2 < 9$

*In Probs. 19 and 20 find the points of intersection of the line and the circle.*

**19**  $y = x - 2$          **20**  $x + y - 1 = 0$

$x^2 + (y + 2)^2 = 32$          $(x - 1)^2 + (y - 1)^2 = 1$

## 4.7   CONICS

The early Greek mathematicians investigated extensively the various curves obtained by cutting a right circular cone with a plane.  These curves, called *conic sections* or *conics* (Fig. 4.28), have simple defini-

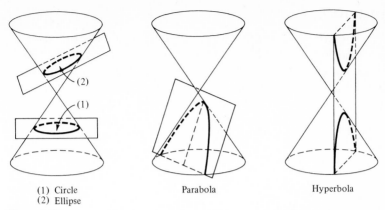

<table>
<tr><td>(1) Circle<br>(2) Ellipse</td><td>Parabola</td><td>Hyperbola</td></tr>
</table>

**Figure 4.28**

tions not involving directly the idea of a section of a cone. The language of these definitions is simplified by the use of the word *locus*, from Latin meaning *place* or *position*. Basically a locus is just a set of places or *points*. As you read the text material in this section, be sure to make a careful study of Figs. 4.28 to 4.39, because they contain many new ideas and definitions of certain important points, lines, and line segments associated with conics. Here are definitions of the conics.

### DEFINITION 4.1

The *circle* is the locus of points P such that P is at a fixed distance (r) from a fixed point called the center. We have already treated the circle; it is a special case of the conic called an *ellipse*. ◀

### DEFINITION 4.2

The *ellipse* is the locus of points P such that the sum of the distances from P to two fixed points F' and F is a constant. The two fixed points are called *foci* (Fig. 4.29). ◀

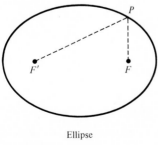

Ellipse

$PF' + PF = \text{constant}$

**Figure 4.29**

## DEFINITION 4.3

The *hyperbola* is the locus of points P such that the absolute value of the difference of the distances from P to two fixed points F' and F is a constant. The two fixed points are called *foci* (Fig. 4.30). ◀

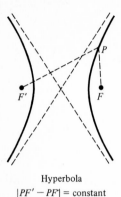

Hyperbola
$|PF' - PF| = \text{constant}$

**Figure 4.30**

## DEFINITION 4.4

The *parabola* is the locus of points P such that the distance from P to a fixed point F is equal to its distance from a fixed line DD'. The fixed point is called the *focus*; the fixed line is called the *directrix* (Fig. 4.31). ◀

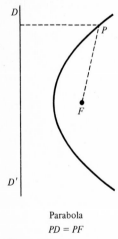

Parabola
$PD = PF$

**Figure 4.31**

    The equation of any curve depends on how the axes and units are chosen. To get the simplest equation for the ellipse we take $F'(-c, 0)$

and $F(c, 0)$ as the foci and take $2a$ as the constant in the definition (Fig. 4.32). From the definition we get:

$$PF' + PF = 2a$$

or
$$\sqrt{(x + c)^2 + y^2} + \sqrt{(x - c)^2 + y^2} = 2a$$

The simplification of this equation is a good exercise in algebra. Transpose the second radical and square.

$$\sqrt{(x + c)^2 + y^2} = 2a - \sqrt{(x - c)^2 + y^2}$$
$$x^2 + 2cx + c^2 + y^2 = 4a^2 - 4a\sqrt{(x - c)^2 + y^2} + x^2 - 2cx + c^2 + y^2$$

This simplifies to

$$cx - a^2 = -a\sqrt{(x - c)^2 + y^2}$$

Square again.

$$c^2x^2 - 2a^2cx + a^4 = a^2(x^2 - 2cx + c^2 + y^2)$$

which reduces to

$$(a^2 - c^2)x^2 + a^2y^2 = a^2(a^2 - c^2) \tag{1}$$

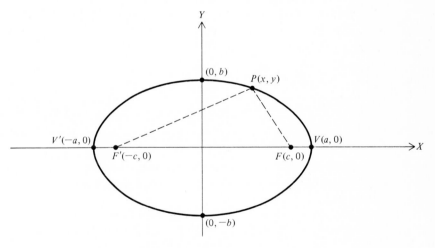

Ellipse: $PF' + PF = 2a$

Equation: $\dfrac{x^2}{a^2} + \dfrac{y^2}{b^2} = 1$

Vertices: $V'(-a, 0)$, $V(a, 0)$
Semimajor axis: $a \ (= V'V/2)$
Semiminor axis: $b$
Where $a^2 - c^2 = b^2$, the foci
are $F'(-c, 0)$, $F(c, 0)$

**Figure 4.32**

Since $a > c$, $a^2 - c^2$ is a positive number, and we set $a^2 - c^2 = b^2$. Equation (1) can therefore be written in the form $b^2x^2 + a^2y^2 = a^2b^2$ and, when this is divided by $a^2b^2$, we get the final and simplest form of the equation of the ellipse:

$$\frac{x^2}{a^2} + \frac{y^2}{b^2} = 1 \qquad \text{equation of ellipse} \qquad (2)$$

The details of the graphs of the following equations are shown in Fig. 4.33.

♦ **Illustration 4.14**

**a** $\dfrac{x^2}{4} + \dfrac{y^2}{1} = 1$ Since $a^2 - b^2 = c^2 = 3$, the foci are $F'(-\sqrt{3}, 0)$, $F(\sqrt{3}, 0)$. The vertices are $V'(-2, 0)$, $V(2, 0)$ (Fig. 4.33a).

**b** $\dfrac{x^2}{25} + \dfrac{y^2}{9} = 1$ Since $a^2 - b^2 = c^2 = 16$, the foci are $F'(-4, 0)$, $F(4, 0)$. The vertices are $V'(-5, 0)$, $V(5, 0)$ (Fig. 4.33b).

**c** $\dfrac{x^2}{10} + \dfrac{y^2}{8} = 1$ Since $a^2 - b^2 = c^2 = 2$, the foci are $F'(-\sqrt{2}, 0)$, $F(\sqrt{2}, 0)$. The vertices are $V'(-\sqrt{10}, 0)$, $V(\sqrt{10}, 0)$ (Fig. 4.33c).

**d** $\dfrac{x^2}{1} + \dfrac{y^2}{4} = 1$ Since $b^2 > a^2$, the semimajor axis is along the $Y$ axis instead of the $X$ axis. $b^2 - a^2 = c^2 = 3$ and the foci are $F'(0, -\sqrt{3})$, $F(0, \sqrt{3})$. The vertices are $V'(0, -2)$, $V(0, 2)$ (Fig. 4.33d). ♦

Some scientific applications of ellipses are:

1   Orbit of a planet (sun at one focus) (Fig. 4.34).
2   Orbits of planetary moons, binary stars, some comets, artificial satellites.
3   Elliptic gears for certain machine tools.
4   Elliptic lathe chuck for turning elliptic cylinders.
5   Many scientific formulas are equations whose graphs are ellipses.
6   Focal property: A ray emanating at one focus is reflected to the other (Fig. 4.35).

To get the simplest equation for the hyperbola we take $F'(-c, 0)$ and $F(c, 0)$ as the foci and $2a$ as the constant in the definition, as in

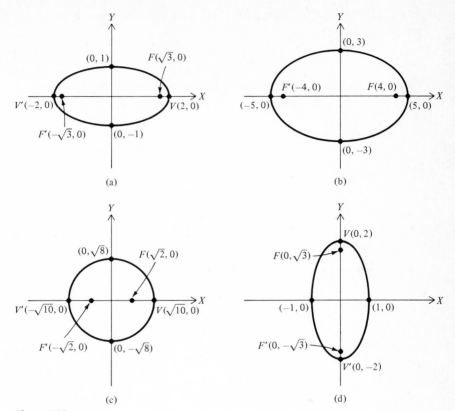

(a)                          (b)

(c)                          (d)

**Figure 4.33**

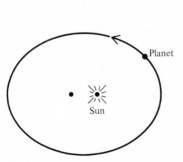

**Figure 4.34**

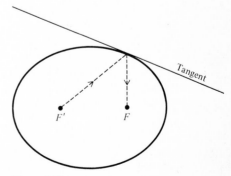

**Figure 4.35**

the case of the ellipse.  But for the hyperbola (Fig. 4.36) we have

$$|PF' - PF| = 2a$$

or      $$\sqrt{(x + c)^2 + y^2} - \sqrt{(x - c)^2 + y^2} = \pm 2a$$

Simplification of this, as in the case of the ellipse, yields

$$(a^2 - c^2)x^2 + a^2y^2 = a^2(a^2 - c^2) \tag{3}$$

and this is exactly the same equation as equation (1) for the ellipse.

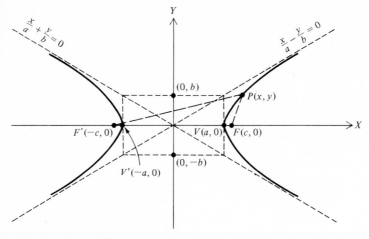

Hyperbola:  $|PF' - PF| = 2a$

Equation: $\dfrac{x^2}{a^2} - \dfrac{y^2}{b^2} = 1$

Vertices:  $V'(-a, 0), V(a, 0)$
Transverse axis is the length $V'V = 2a$
Conjugate axis is the length $2b$
Where $a^2 - c^2 = -b^2$, the foci are $F'(-c, 0), F(c, 0)$

The two lines $\dfrac{x}{a} \pm \dfrac{y}{b} = 0$ are called asymptotes. They are not
part of the graph, but their presence helps in graphing the hyperbola

**Figure 4.36**

But now $a^2 - c^2 < 0$, so we set $a^2 - c^2 = -b^2$ and thus (3) becomes
$-b^2x^2 + a^2y^2 = -a^2b^2$ or

$$\frac{x^2}{a^2} - \frac{y^2}{b^2} = 1 \qquad \text{equation of hyperbola} \tag{4}$$

If we take the left-hand portion of (4) above and set it equal to 0,
we get

$$\frac{x^2}{a^2} - \frac{y^2}{b^2} = 0$$

and this factors into

$$\left(\frac{x}{a} - \frac{y}{b}\right)\left(\frac{x}{a} + \frac{y}{b}\right) = 0$$

The equations $x/a - y/b = 0$ and $x/a + y/b = 0$ are equations of a pair of lines, called *asymptotes*, with the following property: The vertical distance between hyperbola and asymptote gets smaller as $|x|$ gets larger. Check this in Figs. 4.36 and 4.37. The asymptotes should always be sketched in first, because they serve as guidelines in graphing a hyperbola even though they are not part of the hyperbola.

The details of the graphs of the following equations are shown in Fig. 4.37.

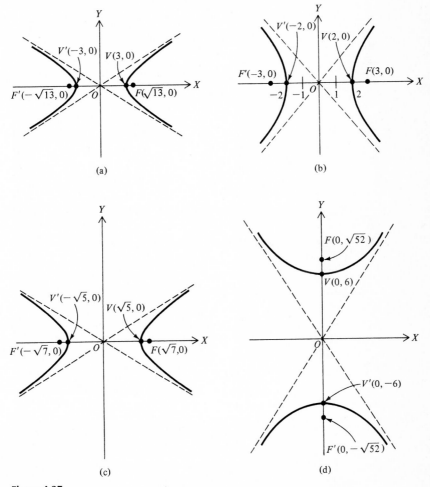

(a)

(b)

(c)

(d)

**Figure 4.37**

♦ **Illustration 4.15**

a  $\dfrac{x^2}{9} - \dfrac{y^2}{4} = 1$    Since $a^2 + b^2 = c^2 = 13$, the foci are $F'(-\sqrt{13}, 0)$ and $F(\sqrt{13}, 0)$. The asymptotes are $x/3 \pm y/2 = 0$, or $2x + 3y = 0$ and $2x - 3y = 0$. The vertices are $V'(-3, 0)$, $V(3, 0)$ (Fig. 4.37a).

b  $\dfrac{x^2}{4} - \dfrac{y^2}{5} = 1$    Since $a^2 + b^2 = c^2 = 9$, the foci are $F'(-3, 0)$ and $F(3, 0)$. The asymptotes are $x/2 \pm y/\sqrt{5} = 0$, or $\sqrt{5}x + 2y = 0$ and $\sqrt{5}x - 2y = 0$. The vertices are $V'(-2, 0)$, $V(2, 0)$ (Fig. 4.37b).

c  $\dfrac{x^2}{5} - \dfrac{y^2}{2} = 1$    Since $a^2 + b^2 = c^2 = 7$, the foci are $F'(-\sqrt{7}, 0)$ and $F(\sqrt{7}, 0)$. The asymptotes are $x/\sqrt{5} \pm y/\sqrt{2} = 0$. The vertices are $V'(-\sqrt{5}, 0)$, $V(\sqrt{5}, 0)$ (Fig. 4.37c).

d  $-\dfrac{x^2}{16} + \dfrac{y^2}{36} = 1$    The transverse axis is now on the $Y$ axis. The foci are on the $Y$ axis. Since $a^2 + b^2 = c^2 = 52$, the foci are $F'(0, -\sqrt{52})$ and $F(0, \sqrt{52})$. The asymptotes are $x/4 \pm y/6 = 0$ or $3x \pm 2y = 0$. The vertices are $V'(0, -6)$, $V(0, 6)$ (Fig. 4.37d). ♦

Some scientific applications of hyperbolas are:

1    Used in the construction of certain telescopic lenses.
2    Some comets trace hyperbolas.
3    Formulas taken from the field of the physical and social sciences are often of the hyperbolic type.

To get the simplest equation for the parabola we place the fixed point at $(p, 0)$ and use as directrix the fixed line $x = -p$ (Fig. 4.38). The equation becomes

$$\sqrt{(x - p)^2 + y^2} = x + p$$

which reduces to

$$y^2 = 4px \qquad \text{equation of parabola}$$

The details of the graphs of the following equations are shown in Fig. 4.39.

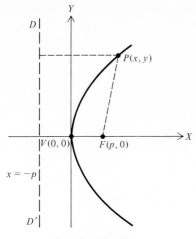

Parabola: $PD = PF$
Equation: $y^2 = 4px$
Vertex: $V(0, 0)$
Focus: $F(p, 0)$
Directrix: $x = -p$

**Figure 4.38**

♦   **Illustration 4.16**

**a**   $y^2 = 8x$        Since $y^2 = 4(2)x$, the focus is $F(2, 0)$.   Directrix, $x = -2$ (Fig. 4.39a).

**b**   $y^2 = 16x$       Since $y^2 = 4(4)x$, the focus is $F(4, 0)$.   Directrix, $x = -4$ (Fig. 4.39b).

**c**   $y^2 = -8x$       Parabola opens up on the negative portion of the X axis, $x \leq 0$.   Focus, $F(-2, 0)$.   Directrix, $x = 2$ (Fig. 4.39c).

**d**   $x^2 = 4y$        Axes interchange roles.   Focus, $F(0, 1)$. Directrix, $y = -1$ (Fig. 4.39d).   ◀

Some applications of parabolas are:

**1**   Path of a projectile, baseball, etc. (neglecting air resistance), with focus at the center of the earth.
**2**   Cable of a suspension bridge (uniformly loaded along the bridge).
**3**   Parabolic reflector [the surface generated by revolving a pa-rabola about its axis has the property that each light ray coming in

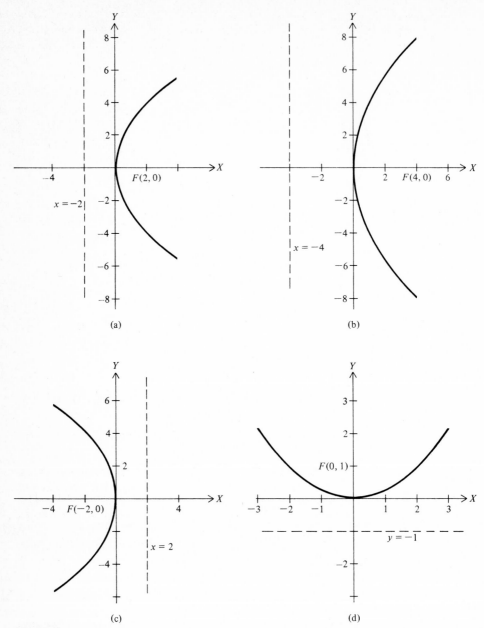

(a)

(b)

(c)

(d)

**Figure 4.39**

parallel to the axis will be reflected to (through) the focus]. Basic principle for flashlights, headlights, searchlights, radio telescopes, etc.

**4**   Graphs of many equations in physics, in econometrics, and in the social sciences are parabolas.

There are other and more complicated equations for conics when they have different orientations with respect to the coordinate axes. Figure 4.40 is the graph of the parabola whose equation is $x^2 + 2xy + y^2 - 2x - 2y + 4 = 0$. The most general equation for a conic is

$$Ax^2 + Bxy + Cy^2 + Dx + Ey + F = 0$$

where not all $A$, $B$, $C$ are 0 because, if they were all 0, the equation would reduce to a linear one whose graph is a line. Note that, if $A = B = F = 0$, $C = 1$, and $D = -4p$, we recover the case of $y^2 = 4px$ discussed above.

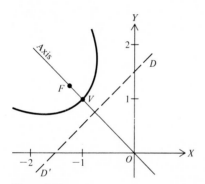

**Figure 4.40**

Another special case is the one where $B = C = 0$ and $E \neq 0$. The equation becomes $Ax^2 + Dx + Ey + F = 0$, and this can be written in the form $-Ey = Ax^2 + Dx + F$, or $y = -Ax^2/E - Dx/E - F/E$. Since $-A/E$, $-D/E$, and $-F/E$ are just real numbers, we can write the equation in the familiar form of Sec. 2.3, namely

$$y = ax^2 + bx + c \qquad (5)$$

Although we do not prove it here, the graph is a parabola. As we saw in Sec. 2.3, the solutions of $ax^2 + bx + c = 0$ are

$$x_1 = \frac{-b + \sqrt{b^2 - 4ac}}{2a} \qquad x_2 = \frac{-b - \sqrt{b^2 - 4ac}}{2a}$$

If either of these two numbers is substituted into equation (5), the right-hand side becomes 0, so that $y = 0$.  The points with coordinates $(x_1, 0)$ and $(x_2, 0)$ lie on the $X$ axis and are the points where the parabola crosses the $X$ axis.  All this happens provided $b^2 - 4ac > 0$.  When $b^2 - 4ac = 0$, $x_1 = x_2$, the parabola has only one point in common with the $X$ axis, and we cannot speak of the parabola crossing the $X$ axis. The parabola is, in fact, tangent to the $X$ axis at the common point. And when $b^2 - 4ac < 0$, $x_1$ and $x_2$ are not real numbers and the parabola has no point in common with the $X$ axis.

◆ **Illustration 4.17**

Sketch the graph of $y = x^2 - x - 2$.  Since $b^2 - 4ac = 1 - (4)(1)(-2) = 9 > 0$, the parabola will cross the $X$ axis at the points where $x^2 - x - 2 = 0$.  This factors readily into $(x - 2)(x + 1) = 0$, so that either $x - 2 = 0$ or $x + 1 = 0$, and so we immediately have the crossing points: $(2, 0)$ and $(-1, 0)$.  We calculate a few other points on the parabola, $(0, -2)$, $(1, -2)$, $(3, 4)$, $(-2, 4)$, $(\frac{1}{2}, -\frac{9}{4})$, and sketch Fig. 4.41.  ◀

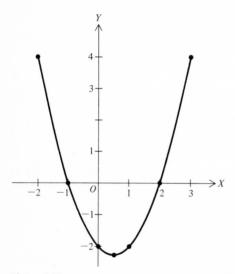

**Figure 4.41**

◆ **Illustration 4.18**

Sketch the graph of $y = x^2 - 6x + 9$.  Here $b^2 - 4ac = (-6)^2 - 4(1)(9) = 0$, and $x_1 = x_2 = -b/2a = \frac{6}{2} = 3$.  The point of tangency is $(3, 0)$.  It is simple to find other points by using the right-hand side of the identity

$x^2 - 6x + 9 = (x - 3)^2$. Some other points are $(0, 9)$, $(1, 4)$, $(2, 1)$, $(4, 1)$, $(5, 4)$, $(6, 9)$ (Fig. 4.42). ◀

▶ **Illustration 4.19**

Sketch the graph of $y = x^2 + x + 1$. This parabola does not have a point in common with the $X$ axis, since

$$\frac{-b \pm \sqrt{b^2 - 4ac}}{2a} = \frac{-1 \pm \sqrt{1 - 4}}{2}$$

and these are not real numbers. Some points on the graph are $(0, 1)$, $(1, 3)$, $(2, 7)$, $(-\frac{1}{2}, \frac{3}{4})$, $(-1, 1)$, $(-2, 3)$, $(-3, 7)$ (Fig. 4.43). ◀

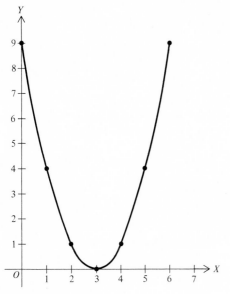

**Figure 4.42**

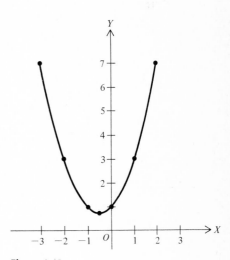

**Figure 4.43**

## PROBLEMS 4.7

In Probs. 1 to 12 find the coordinates of the focus (foci), the vertex (vertices), the equation of the directrix, and/or the equations of the asymptotes, whichever apply. Sketch the conic.

**1** $\dfrac{x^2}{25} + \dfrac{y^2}{16} = 1$     **2** $\dfrac{x^2}{16} + \dfrac{y^2}{9} = 1$     **3** $\dfrac{x^2}{144} + \dfrac{y^2}{16} = 1$

**4** $\dfrac{x^2}{9} + \dfrac{y^2}{25} = 1$     **5** $\dfrac{x^2}{25} - \dfrac{y^2}{9} = 1$     **6** $\dfrac{x^2}{16} - \dfrac{y^2}{144} = 1$

**7** $\dfrac{x^2}{49} - \dfrac{y^2}{16} = 1$     **8** $-\dfrac{x^2}{4} + \dfrac{y^2}{1} = 1$     **9** $y^2 = 12x$

**10** $y^2 = x$     **11** $y^2 = -20x$     **12** $x^2 = 2y$

*In Probs. 13 to 16, sketch the conic in the indicated interval.*

**13** $y = x(x + 1), -3 \le x \le 2$     **14** $y = 2x(x - 1), -1 \le x \le 2$

**15** $y = x - x^2, -1 \le x \le 2$     **16** $y = 4 - x^2, -3 \le x \le 3$

## 4.8    LINEAR PROGRAMMING

We shall now apply our work on linear inequalities to some examples in a branch of mathematics called linear programming. Although it is a new subject, having been created in the 1940s, it is now quite extensive in theory and is applicable to a very wide range of problems. We start with a simple example.

A small furniture manufacturer makes two types of chairs, regulars and rockers. Due to limitations of equipment and manpower she can make at most 14 regulars and at most 12 rockers per day. Further, her combined daily production cannot exceed 20 chairs. She can sell all she can make with a profit of $2 per regular and $4 per rocker. How many chairs of each kind should she make in order to realize a daily maximum profit?

Let $x$ be the number of regulars and $y$ the number of rockers. Clearly $x$ and $y$ satisfy certain inequalities: $x$ and $y$ are nonnegative in the first place, and $x$ cannot exceed 14 and $y$ cannot exceed 12. We write this information down in double-inequality form:

$$0 \le x \le 14 \tag{1}$$

$$0 \le y \le 12 \tag{2}$$

We plot lines $x = 14$ and $y = 12$ in Fig. 4.44 and note that permissible values of $x$ and $y$ are confined to the rectangle $OABC$. Also the total daily output cannot exceed 20, and this leads to the inequality

$$x + y \le 20 \tag{3}$$

and we plot the line $x + y = 20$ in Fig. 4.44. Testing (3) at the origin shows that the solution set lies below the line, and this further restricts $x$ and $y$ to lie in the region $OADEC$. This region is called the *feasible region*: The solution set of the simultaneous inequalities (1), (2), and

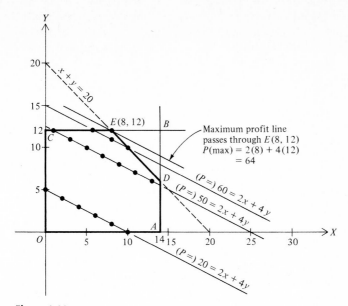

**Figure 4.44**

(3) is the set of number pairs $(x, y)$ lying in and on $OADEC$. It is easy to find the coordinates of $E(8, 12)$ and $D(14, 6)$.

So far this is nothing new; we did similar problems in Sec. 4.4. But now comes the matter of the profit $P$. Since she makes \$2 on each regular, she makes $2x$ on $x$ of them, and similarly $4y$ on the rockers. Therefore the total profit equation is

$$P = 2x + 4y \qquad (4)$$

and $P$ will vary with $x$ and $y$. We know that $x$ and $y$ must lie in or on $OADEC$, and so we wonder what $x$ and $y$ would yield a profit of, say, \$20. The profit equation is then

$$20 = 2x + 4y \qquad (5)$$

We plot this line in Fig. 4.44 and readily determine that she could make a profit of \$20 by making any one of several combinations of regulars and rockers as exhibited in Table 4.6 and plotted as isolated points on the line $(P=) \; 20 = 2x + 4y$. (In this problem $x$ and $y$ have to be whole numbers.)

**Table 4.6**

| x regulars | 0 | 2 | 4 | 6 | 8 | 10 |
|---|---|---|---|---|---|---|
| y rockers | 5 | 4 | 3 | 2 | 1 | 0 |

Figure 4.44 also shows two more profit lines, one for $50 profit (for which there are seven combinations of regulars and rockers, Table 4.7) and one for $60 profit (for which there are only three combinations, Table 4.8).

**Table 4.7**

| x | 1 | 3 | 5 | 7 | 9 | 11 | 13 |
|---|---|---|---|---|---|----|----|
| y | 12 | 11 | 10 | 9 | 8 | 7 | 6 |

**Table 4.8**

| x | 6 | 8 | 10 |
|---|---|---|----|
| y | 12 | 11 | 10 |

There are two important things to note about the profit lines:

**1** The lines are parallel with slope $-\frac{1}{2}$.
**2** The greater the profit the further away from the origin is the profit line.

The greatest distance a profit line $P = 2x + 4y$ could be from the origin would be that of the line passing through E, because any further increase in $P$ would produce a line having no point in common with the feasible region. Therefore the greatest profit $P$ is obtained when $P = 2(8) + 4(12)$ and is $P = 64$, which is obtained by substituting into the profit equation the coordinates of point $E(8, 12)$. For this maximum profit she should make 8 regulars and 12 rockers.

To modify the problem a bit suppose that the profit on a regular is $4 and only $2 on a rocker. The new profit line is $P = 4x + 2y$ with slope $-2$. A few such lines are plotted in Fig. 4.45. As the profits

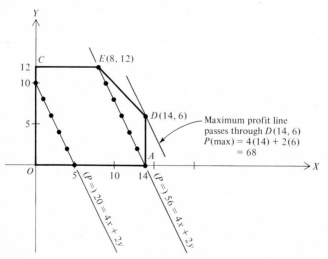

**Figure 4.45**

$P$ increase, the lines move away from the origin as before and still cut the feasible region after sweeping through $E$. The profit line passing through $D$ now gives maximum profit: $P(\text{max}) = 4(14) + 2(6) = 68$.

Finally, note what happens if the profit is the same on the two kinds of chairs, say, \$2. Then $P = 2x + 2y$ and these lines have slope $-1$ (Fig. 4.46). This is the same slope as that of the line $ED$, or $x + y = 20$. In this case the greatest profit line will coincide with line $ED$, and so there are several combinations of chairs each giving the same maximum profit (Table 4.9). In each case $P = 2x + 2y = 2(x + y) = 40$.

**Table 4.9**

| x | 8 | 9 | 10 | 11 | 12 | 13 | 14 |
|---|---|---|----|----|----|----|----|
| y | 12 | 11 | 10 | 9 | 8 | 7 | 6 |

In this problem the data have been specially chosen so that answers come out in terms of whole numbers, but it would not be of much concern if, for a certain profit, the number of regular chairs turned out to be 6.7. It would be similar to the case of the average number of children in the U.S. families, which is something like 2.4.

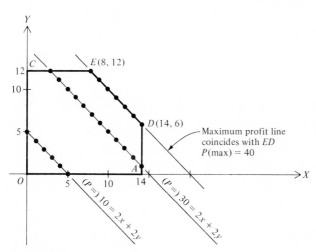

**Figure 4.46**

▶ **Illustration 4.20**

Let us work through another example. A man needs 0.38 gram of a special food supplement $R$ and 0.42 gram of another food supplement $S$ daily. These cannot be separated easily but come naturally in mixtures $M$ and $N$. Mixture $M$ has 10 percent $R$, 40 percent $S$, and

50 percent inert matter, while mixture $N$ has 30 percent $R$, 10 percent $S$, and 60 percent inert matter. The costs of the mixtures are \$2 per gram and \$3 per gram for $M$ and $N$, respectively. How many grams of $M$ and $N$ should he buy to minimize his cost for a 10-day supply?

Let $x$ be the number of grams of $M$ bought and $y$ the number of grams of $N$ bought. Both $x$ and $y$ are nonnegative, so that the first restrictions are

$$0 \leq x \tag{1}$$
$$0 \leq y \tag{2}$$

If $x$ grams of $M$ are bought, $0.1x$ is the number of grams of $R$ present and if $y$ grams of $N$ are bought, $0.3y$ is the additional number of grams of $R$ available. Therefore, for a 10-day supply of $R$ we have

$$0.1x + 0.3y \geq 3.8 \tag{3}$$

Similarly, the inequality for $S$ is

$$0.4x + 0.1y \geq 4.2 \tag{4}$$

It is best to multiply (3) and (4) by 10.

$$x + 3y \geq 38 \tag{3'}$$
$$4x + y \geq 42 \tag{4'}$$

The two lines ($AB$ and $CD$, Fig. 4.47) intersect at point $E(8, 10)$. Testing (3') and (4') at $(0, 0)$ we find that the feasible region is the portion of the first quadrant *above* these two lines.

The question now is what $x$ and what $y$ will give him the supplements he needs at least cost. Each cost equation is of the form $C = 2x + 3y$, and each has slope $-\frac{2}{3}$. Figure 4.47 shows the cost lines for $C = 80$ and $C = 60$. It should be clear that the cost decreases as the cost line moves toward the origin. The minimum cost line therefore passes through $E(8, 10)$, and the minimum cost is $C(\min) = 2(8) + 3(10) = 46$. ◀

*Remark 1*   Both the algebra and the geometry are easy in linear programming problems with just two variables. The geometry (of space) is very much harder for three variables and is nonexistent for more than three variables. Problems in a large number of variables are handled by algebra and computers.

*Remark 2*   General theories of linear programming show that maximum (or minimum) solutions always occur, if unique, at a vertex (corner) of the feasible region, as in these two illustrations.

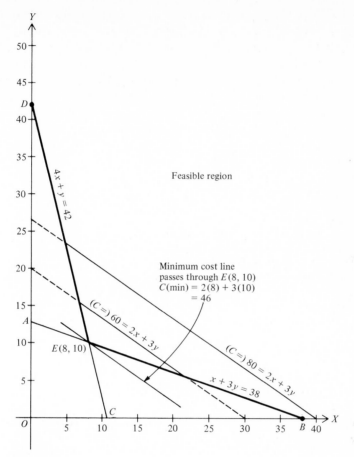

Figure 4.47

## PROBLEMS 4.8

*In each of the following problems: (a) Write down all inequalities. (b) Plot the lines determined by the equalities. (c) Show clearly the feasible region. (d) Determine the maximum (minimum) called for.*

**1** A small warehouse has a maximum of 100 cubic feet storage space for storing two types of merchandise, A and B. Type A comes in 2-cubic-foot cartons costing $3 apiece, and B comes in 1-cubic-foot cartons costing $4 apiece. The owner can spend at most $300. After storage for a year each type brings a profit of $10. How many A's and B's should he store for maximum profit?

**2** To raise money for a charity, there is to be an hour of entertainment for the community. The community will pay for the entertainment and donate an equal amount to the charity. The hour's entertainment will be a variety show for $x$ minutes and some music for $y$ minutes, but $x + y$ is not to exceed 55 minutes, because the manager says he wants at least 5 minutes to make a speech. And he will talk any extra time if the show and music do not use the full 55 minutes. The variety show is to run for at least 25 minutes but is not to exceed 50 minutes. The music is not to exceed 20 minutes. The variety show will cost $40 per minute and the music $10 per minute. What allocations should be made for the show and the music to yield a maximum donation?

**3** A shoe manufacturer is geared to a daily production of at most 55 pairs of shoes and 40 pairs of boots. She requires 1 square foot of leather to make a pair of shoes and 2 square feet to make a pair of boots. Leather is in short supply, and she has only 100 square feet available daily. She ships 3 pairs of shoes and 1 pair of boots to her retail outlet in a shipping box, but these also are in short supply and she cannot use more than 180 boxes per day. If her profit is $2 per pair of shoes and $3 per pair of boots, how many pairs of each should she produce daily for maximum profit?

**4** A company makes $x$ metal light standards of type $A$ and $y$ standards of type $B$. These are processed on three machines, W (welding), M (milling), and D (drilling). The table below gives the processing time (hours) for each type and for each machining operation.

|   | W | M | D |
|---|---|---|---|
| $A$ | 3 | 6 | 9 |
| $B$ | 12 | 6 | 3 |

Machine W can be used at most 120 hours per week, machine M, 96 hours, and machine D, 108 hours. The company makes a unit of money profit on each type of standard, the unit being of no importance here. How many standards of type $A$ and $B$ should the company make per week for maximum profit?

**5** To provide sufficient nitrogen (N), phosphorus (P), and lime (L) for her 400-acre farm, Mrs. Hill buys $x$ bags of fertilizer $A$ and $y$ bags of fertilizer $B$. The data are in the table:

|   | N | P | L | Cost per bag |
|---|---|---|---|---|
| $A$ | 6 | 2 | 8 | $ 5 |
| $B$ | 6 | 9 | 15 | $10 |
| Minimum need per acre | 108 | 85 | 235 | |

Find the values of $x$ and $y$ which will meet at least the minimum need for fertilizer and for which the cost per acre is a minimum.

**6** Housing must be provided for at least 435 persons at a remote mine. Buildings must be of two types, $A$ and $B$. The data are:

| Type of house | Units of lumber | Units of concrete | Capacity (persons) | Cost | No. of houses |
|---|---|---|---|---|---|
| A | 10 | 7 | 12 | $ 5,000 | x |
| B | 5 | 25 | 25 | $12,000 | y |

At least 125 units of lumber and 285 units of concrete must be used. Find the values of x and y for which the cost of construction is a minimum.

# 5

# POLYNOMIAL
# FUNCTIONS

## 5.1  INTRODUCTION AND DEFINITION

The most useful elementary functions that you will study are the polynomial functions or, simply, polynomials.  Not only do they arise
directly in many different applications, but they are also useful in
analyzing other, more complicated, functions.

In Chap. 3 brief mention was made of polynomials.  In particular,
you gave some attention to linear and quadratic functions, which are
polynomials.

### ◆  Illustration 5.1

Such linear functions as

$$f(x) = 2x + 1$$
$$g(t) = 0.3t - \tfrac{1}{2}$$
$$h(v) = \sqrt{2}v + \pi$$

are all polynomial functions.  ◆

◆ **Illustration 5.2**

Quadratic functions such as

$$f(x) = 3x^2 - 7x + \sqrt{7}$$
$$g(t) = t^2$$
$$h(v) = 4v^2 - 2$$

are polynomial functions. ◀

◆ **Illustration 5.3**

$f(x) = x^3 - 7x + 2$ is a *cubic* polynomial or polynomial of *degree* 3

$$g(t) = t^{23} - 17t^{15} + \sqrt{2}t^4 + 13t$$

is a polynomial of degree 23.

$$h(x) = 2$$

is a polynomial of degree 0. (Note that $2 = 2x^0$.) ◀

◆ **Illustration 5.4**

$$f(x) = \frac{1}{x}$$
$$g(t) = 2\sqrt{t} + 3t$$
$$h(r) = 2^r$$

are not polynomial functions. ◀

Before reading the definition given below, look back over these examples to determine the essential features of a polynomial function.

### DEFINITION 5.1

**a** Any function that can be written in the form

$$P(x) = a_n x^n + a_{n-1} x^{n-1} + \cdots + a_1 x + a_0$$

is a *polynomial function*, where $a_n, a_{n-1}, \ldots, a_1, a_0$ are real numbers, and where $n$ is a whole number.

**b** $a_n, a_{n-1}, \ldots, a_1, a_0$ are called the *coefficients* of the polynomial function.

**c** The degree of the polynomial function is the largest exponent of $x$ that appears. ◀

Your analysis of polynomial functions in this chapter will focus on two important and interrelated concerns:

the *zeros* of polynomials, i.e., the values of x which make P(x) equal to 0.

the graphs of polynomials.

First, we give you a few problems to make sure you know what the words and symbols used so far mean.

## PROBLEMS 5.1

**1** Functions relate sets of numbers. If $P(x) = x^3 - 2x^2 + 3x - \sqrt{2}$, find the number P(x) which is related to each of the following values of x.

    **a** $x = 1$      **b** $x = 0$      **c** $x = \sqrt{2}$

    **d** $x = \sqrt[3]{2}$      **e** $x = -1$      **f** $x = 2$

**2** If $P(x) = x^3 + 4x^2 - 2x + 5$, find

    **a** $P(-2)$      **b** $P(0)$      **c** $P(5)$

    **d** $P(\sqrt{3})$      **e** $P(1)$      **f** $P(\frac{1}{2})$

**3** Determine which of the following are polynomials. In the case of a polynomial determine its degree.

    **a** $P(x) = x^{10}$      **b** $P(x) = x^{1/2}$

    **c** $P(x) = \dfrac{x^2 + 1}{x + 1}$      **d** $P(x) = x^2 + 2x + 1$

    **e** $P(x) = \sin x$      **f** $P(x) = -2x^{1000} + \sqrt{2}x^{755} + 3x^2 - \pi$

    **g** $P(x) = 0$

**4** Identify the polynomials and their degrees from among the following functions.

    **a** $P(x) = \sqrt[3]{x}$      **b** $P(x) = 3^x + 2x + 1$

    **c** $P(x) = 3x - \frac{1}{2}$      **d** $P(x) = \dfrac{x - 1}{x + 1}$

    **e** $P(x) = \dfrac{1}{x}$      **f** $P(x) = -7x^{121} + 3x^2 + 4$

    **g** $P(x) = \pi$

**5** A bicycle manufacturer suspects that the relationship between the price of his best bicycle and his company's profit from that line of bicycles is linear; that is, $P(x) = ax + b$ expresses the relationship between the profit P and the price x for some choice of a and b. Assuming that the manufacturer is correct and that the company makes a profit of $100,000 when the price is $100, and of $125,000 when the price is $115,

    **a** Find the coefficients a and b for the profit function.

    **b** Determine what the profit would be if the price were $130.

**c** Then analyze the profit function $P(x)$ to see if it would make sense for profits to behave that way.

**6** The following temperatures were observed on consecutive days in January:

| Date | Temperature |
| --- | --- |
| 12 | $5°$ |
| 13 | $0°$ |
| 14 | $10°$ |

It appears that the temperatures for the month followed a quadratic (second-degree polynomial) curve; that is, $T(x) = ax^2 + bx + c$, where $T$ is the temperature and $x$ is the day of the month. If this is true,

**a** Find the coefficients $a$, $b$, $c$ of the temperature function.
**b** Find the temperature on January 3.

## 5.2   ZEROS BY FORMULA

We are going to be analyzing polynomials to find their zeros. In a few cases this analysis is simple and can be done in general using a formula. We briefly present those cases here.

### Polynomials of Degree 0
A zero-degree polynomial is a constant function

$$P(x) = a \qquad \text{for all } x$$

Such a polynomial either has no zeros (if $a \neq 0$) or has an infinite number of zeros (if $a = 0$). For example, the function $P(x) = 2$ has no zeros.

### Polynomials of Degree 1
Polynomials of degeee 1 all have exactly one zero which can be found easily as follows.

Suppose $P(x) = ax + b$. If $P(x) = 0$,

$$ax + b = 0$$
$$ax = -b$$
$$x = -\frac{b}{a}$$

For example, if $P(x) = -2x + 3$, $x = \frac{3}{2}$ is the zero of $P(x)$.

### Polynomials of Degree 2
The situation for polynomials of degree 2 is trickier.

In your earlier work with the quadratic formula you learned that, if $P(x) = ax^2 + bx + c$, then

$$P(x) = 0 \quad \text{when } x = \frac{-b \pm \sqrt{b^2 - 4ac}}{2a}$$

If $b^2 - 4ac > 0$, then

$$\frac{-b + \sqrt{b^2 - 4ac}}{2a} \quad \text{and} \quad \frac{-b - \sqrt{b^2 - 4ac}}{2a}$$

are two real numbers that are zeros of $P(x)$. If $b^2 - 4ac = 0$, then $-b/2a$ is the only zero of $P(x)$. It is a repeated zero in a sense that will be explained later.

If $b^2 - 4ac < 0$, then $\sqrt{b^2 - 4ac}$ is a complex number and $(-b \pm \sqrt{b^2 - 4ac})/2a$ provides two complex zeros. We will focus our attention on the real zeros of polynomials.

▶ **Illustration 5.5**

$$P(x) = 2x^2 + 4x + 1$$

$$P(x) = 0 \quad \text{if } x = \frac{-4 \pm \sqrt{(4)^2 - 4(2)(1)}}{2(2)} = \frac{-4 \pm \sqrt{8}}{4}$$

Notice that $b^2 - 4ac = 8 > 0$. ◀

▶ **Illustration 5.6**

$$P(x) = 4x^2 - 12x + 9$$

$$P(x) = 0 \quad \text{if } x = \frac{-(-12) \pm \sqrt{(-12)^2 - 4(4)(9)}}{2(4)}$$

$$= \frac{12 \pm \sqrt{0}}{8} = \frac{12}{8} = \frac{3}{2}$$

Notice that $b^2 - 4ac = 0$. ◀

▶ **Illustration 5.7**

$$P(x) = x^2 + x + 1$$

$$P(x) = 0 \quad \text{if } x = \frac{-1 \pm \sqrt{1 - 4(1)(1)}}{2(1)}$$

$$= \frac{-1 \pm \sqrt{-3}}{2}$$

These are complex numbers, since $b^2 - 4ac < 0$. ◀

There is a complicated formula for third-degree (cubic) polynomials and also one for fourth-degree (quartic). But there is no formula for fifth-degree (quintic), and there is no formula for any of higher degree. This amazing fact was proved by the Norwegian mathematician Abel in 1824. Abel's proof followed hundreds of years of effort by mathematicians to find such formulas. In the absence of formulas, we will have to develop some analytic techniques for "tracking down" the zeros of polynomials. Before we get on with that, here are some problems on which you can practice the use of formulas for finding the zeros of polynomials.

## PROBLEMS 5.2

**1**  Find the zero(s) of each of the following polynomials.

   **a**  $P(x) = 3x - \frac{2}{3}$           **b**  $P(x) = \pi - x$

   **c**  $P(x) = x^2 - 3x + 2$      **d**  $P(x) = x^2 - 2x + 1$

   **e**  $P(x) = x^2 + x - 1$

**2**  Find the zero(s) of each of the following polynomials.

   **a**  $P(x) = 4(3x + 2)$        **b**  $P(x) = 2\sqrt{2}x - 1$

   **c**  $P(x) = x^2 - 2$            **d**  $P(x) = x^2 - 5x + 6$

   **e**  $P(x) = x^2 + 2x - 4$

**3**  There are some polynomials of degree higher than 2 for which you *can* find the zeros. Find as many zeros of each of the following polynomials as you can.

   **a**  $P(x) = x^{20} - 1$

   **b**  $P(x) = (x - 1)(x - 2)(x - 3)(x - 4)(x - 5)$ (This is a polynomial in factored form.)

   **c**  $P(x) = x^4 - 3x^2 + 2$ (There is a tricky way to treat this like a quadratic.)

**4**  Find as many zeros of each of the following polynomials as you can.

   **a**  $P(x) = 7x^5 - 224$

   **b**  $P(x) = (x - 1)(x^2 - 3x + 2)$ (This is a polynomial in a partially factored form.)

   **c**  $P(x) = 2x^4 - 10x^2 + 8$

## 5.3   OPERATIONS ON POLYNOMIALS

In the next few sections we are going to be developing procedures for finding the zeros of polynomials. In this process we will need to be able to combine polynomials using the operations of addition, subtraction, multiplication, and division. This section introduces you to these four operations on polynomials.

### Addition and Subtraction

The sum and difference of two polynomials are polynomials, as the following illustrations suggest.

▶ **Illustration 5.8**

If $p(x) = 3x^2 + 2x + 4$ and $q(x) = \sqrt{2}x - 7$,

$$\begin{aligned}
p(x) + q(x) &= (3x^2 + 2x + 4) + (\sqrt{2}x - 7) \\
&= 3x^2 + (2 + \sqrt{2})x + (4 - 7) \\
&= 3x^2 + (2 + \sqrt{2})x - 3 \\
p(x) - q(x) &= (3x^2 + 2x + 4) - (\sqrt{2}x - 7) \\
&= 3x^2 + (2 - \sqrt{2})x + [4 - (-7)] \\
&= 3x^2 + (2 - \sqrt{2})x + 11 \quad ◀
\end{aligned}$$

▶ **Illustration 5.9**

If $p(x) = 5x^4 - 3x^2 + x$ and $q(x) = x^3 + 2x^2 - x + 3$,

$$\begin{aligned}
p(x) + q(x) &= 5x^4 + x^3 - x^2 + 3 \\
p(x) - q(x) &= 5x^4 - x^3 - 5x^2 + 2x - 3 \quad ◀
\end{aligned}$$

In general if

$$p(x) = a_n x^n + a_{n-1}x^{n-1} + \cdots + a_1 x + a_0$$

and

$$q(x) = b_n x^n + b_{n-1}x^{n-1} + \cdots + b_1 x + b_0$$

then

$$p(x) \pm q(x) = (a_n \pm b_n)x^n + (a_{n-1} \pm b_{n-1})x^{n-1} + \cdots \\ + (a_1 \pm b_1)x + (a_0 \pm b_0)$$

That is, to add (subtract) polynomials you add (subtract) coefficients of the same powers of the variable.

### Multiplication

The product of polynomials is a polynomial. The procedure for computing that polynomial is a little harder than the one for addition and subtraction. It follows from the rule for multiplying real numbers:

$$(a + b)(c + d) = ac + ad + bc + bd$$

Note that each of $a$ and $b$ is multiplied by each of $c$ and $d$, and then all the products are added.  Work through these illustrations.

▶ **Illustration 5.10**

If $p(x) = 2x + 1$ and $q(x) = -3x + 5$,

$$\begin{aligned}
p(x) \cdot q(x) &= (2x + 1)(-3x + 5) \\
&= (2x)(-3x) + (2x)(5) + (1)(-3x) + (1)(5) \\
&= -6x^2 + 10x - 3x + 5 \\
&= -6x^2 + 7x + 5 \quad ◀
\end{aligned}$$

▶ **Illustration 5.11**

The same problem can be worked using a different format.  This format helps keep track of the powers of $x$.  This is especially helpful with more complex polynomials.

$$
\begin{array}{r}
2x + 1 \\
-3x + 5 \\
\hline
-6x^2 - 3x \quad\;\; \\
+10x + 5 \\
\hline
-6x^2 + 7x + 5 \quad ◀
\end{array}
$$

▶ **Illustration 5.12**

If $p(x) = 7x^3 - 5x^2 + 2x - 6$ and $q(x) = -2x^3 - x + 5$, then $p(x) \cdot q(x)$ can be computed in either of the following ways.

$$\begin{aligned}
p(x) \cdot q(x) &= (7x^3 - 5x^2 + 2x - 6)(-2x^3 - x + 5) \\
&= (7x^3)(-2x^3) + (7x^3)(-x) + (7x^3)(5) \\
&\quad + (-5x^2)(-2x^3) + (-5x^2)(-x) + (-5x^2)(5) \\
&\quad + (2x)(-2x^3) + (2x)(-x) + (2x)(5) \\
&\quad + (-6)(-2x^3) + (-6)(-x) + (-6)(5) \\
&= -14x^6 + 10x^5 - 11x^4 + 52x^3 - 27x^2 + 16x - 30
\end{aligned}$$

$$
\begin{array}{r}
7x^3 - 5x^2 + 2x - 6 \\
-2x^3 - \;\; x + 5 \\
\hline
-14x^6 + 10x^5 - \;\; 4x^4 + 12x^3 \qquad\qquad\qquad\quad\; \\
- \;\; 7x^4 + \;\; 5x^3 - \;\; 2x^2 + \;\; 6x \qquad\quad\; \\
+ 35x^3 - 25x^2 + 10x - 30 \\
\hline
-14x^6 + 10x^5 - 11x^4 + 52x^3 - 27x^2 + 16x - 30 \quad ◀
\end{array}
$$

In general the procedure for multiplying two polynomials is to multiply each term of one by each term of the other and then add the resulting terms, combining coefficients of the same power of x.

### Division

The division of polynomials has a great deal in common with the division of whole numbers. To start with, just as $37 \div 23$ is not a whole number, the quotient of two polynomials may not be a polynomial. Also, just as $a/b = c$ is equivalent to $a = bc$ for $b \neq 0$, $p(x)/q(x) = f(x)$ is equivalent to $p(x) = q(x)f(x)$, where $q(x) \neq 0$. Moreover, the actual procedure for dividing polynomials is not unlike that for long division of whole numbers. Read through the steps of the following illustrations.

#### ▶ Illustration 5.13

Let $p(x) = x^3 + 2x + 4$ and $q(x) = x^2 - 2$. Then $p(x)/q(x)$ is computed as follows.

$$
\begin{array}{r}
x \phantom{xxxxxxxxxxxxx} \\
x^2 - 2 \overline{)x^3 + 0x^2 + 2x + 4} \\
\underline{x^3 \phantom{xxxxx} - 2x} \\
4x + 4
\end{array}
$$

So dividing $x^2 - 2$ into $x^3 + 2x + 4$ results in a quotient of x with remainder $4x + 4$. This can be written as

$$
\frac{x^3 + 2x + 4}{x^2 - 2} = x + \frac{4x + 4}{x^2 - 2} \quad ◀
$$

In general to divide the polynomial $p(x)$ by the polynomial $q(x)$,

**1** Write $p(x)$ out, leaving a space for missing powers of x.
**2** Divide the highest-powered term of $q(x)$ into the highest-powered term of $p(x)$.
**3** Proceed to multiply, subtract, bring down another term of $p(x)$, and repeat the process much as in the division of whole numbers until the degree of the remainder is less than that of $q(x)$.

Go back over the example above to see that these steps were followed. Then read through the following illustration.

**Illustration 5.14**

Let $p(x) = x^3 + 2x^2 - 5$ and $q(x) = 2x - 3$.

$$
\begin{array}{r}
\frac{1}{2}x^2 + \frac{7}{4}x + \frac{21}{8} \\[2pt]
2x - 3 \overline{)\, x^3 + 2x^2 + 0x - 5}
\end{array}
$$

$$x^3 - \tfrac{3}{2}x^2 \qquad \begin{cases} \dfrac{x^3}{2x} = \dfrac{1}{2}x^2 \\[4pt] (2x-3)(\tfrac{1}{2}x^2) = x^3 - \tfrac{3}{2}x^2 \end{cases}$$

$$\tfrac{7}{2}x^2 + 0 \qquad \begin{cases} (x^3 + 2x^2) - (x^3 - \tfrac{3}{2}x^2) = \tfrac{7}{2}x^2, \\[4pt] \text{bring down } 0x = 0 \end{cases}$$

$$\tfrac{7}{2}x^2 - \tfrac{21}{4}x \qquad \begin{cases} \dfrac{\tfrac{7}{2}x^2}{2x} = \dfrac{7}{4}x \\[4pt] (2x-3)(\tfrac{7}{4}x) = \tfrac{7}{2}x^2 - \tfrac{21}{4}x \end{cases}$$

$$\tfrac{21}{4}x - 5 \qquad \begin{cases} (\tfrac{7}{2}x^2 + 0) - (\tfrac{7}{2}x^2 - \tfrac{21}{4}x) = \tfrac{21}{4}x, \\[4pt] \text{bring down } -5 \end{cases}$$

$$\tfrac{21}{4}x - \tfrac{63}{8} \qquad \begin{cases} \dfrac{\tfrac{21}{4}x}{2x} = \dfrac{21}{8} \\[4pt] (2x-3)(\tfrac{21}{8}) = \tfrac{21}{4}x - \tfrac{63}{8} \end{cases}$$

$$\tfrac{23}{8} \qquad \{(\tfrac{21}{4}x - 5) - (\tfrac{21}{4}x - \tfrac{63}{8}) = \tfrac{23}{8}$$

So
$$\frac{x^3 + 2x^2 - 5}{2x - 3} = \frac{1}{2}x^2 + \frac{7}{4}x + \frac{21}{8} + \frac{\frac{23}{8}}{2x - 3}$$

If you want to check the answer, multiply through to see if

$$x^3 + 2x^2 - 5 = (2x - 3)(\tfrac{1}{2}x^2 + \tfrac{7}{4}x + \tfrac{21}{8}) + \tfrac{23}{8}$$

Remember that $p(x)/q(x) = f(x)$ is equivalent to $p(x) = q(x) \cdot f(x)$. ◀

## PROBLEMS 5.3

*For each of the following pairs of polynomials find $p(x) + q(x)$, $p(x) - q(x)$, $p(x) \cdot q(x)$ and $p(x)/q(x)$. In the case of $p(x)/q(x)$, check your answer by multiplying.*

1  $p(x) = x^2 + 2x - 3$, $q(x) = x + 4$
2  $p(x) = 3x^3 + 3x^2 - 4x + 1$, $q(x) = x^2 - 3x + 2$
3  $p(x) = x^4 + x^2 - 1$, $q(x) = 2x^2 + 1$
4  $p(x) = x^5 - 1$, $q(x) = x - 1$
5  $p(x) = x^3 - 2x^2 + 7x - 1$, $q(x) = x + 3$
6  $p(x) = 4x^3 + 2x + 6$, $q(x) = x^2 - 1$
7  $p(x) = x^4 + x + 1$, $q(x) = 2x + 3$
8  $p(x) = x^6 - 1$, $q(x) = x - 1$

## 5.4    GENERAL PROCEDURE FOR FINDING ZEROS

You have seen that there are formulas for finding the zeros of polynomials of degree 1 and 2, and we have indicated that there are such formulas for polynomials of degree 3 and 4. In this section we will introduce some techniques that will help you to locate zeros of polynomials. Your role will not be unlike that of a detective who sets out to track down a missing person. Sometimes the polynomial will give you sufficient clues right off for you to find all of its zeros. Other times your clues may help you get somewhat closer to a zero but not guarantee that you will ever actually find one. While most of the examples worked are polynomials of degree 3 and 4, the techniques apply to polynomials of any degree.

To help in finding zeros of polynomials we will need to learn some general facts about them. In particular we need to know about division of polynomials. In Sec. 5.3 we saw that

$$\frac{x^3 + 2x + 4}{x^2 - 2} = x + \frac{4x + 4}{x^2 - 2}$$

and that

$$\frac{x^3 + 2x^2 - 5}{2x - 3} = \frac{1}{2}x^2 + \frac{7}{4}x + \frac{21}{8} + \frac{\frac{23}{8}}{2x - 3}$$

Both of these expressions are in the form

$$\frac{P(x)}{D(x)} = Q(x) + \frac{R(x)}{D(x)}$$

or

$$P(x) = Q(x)D(x) + R(x)$$

Note also that

The degree of the divisor $D(x)$ is less than that of $P(x)$.
The degree of $Q(x)$ is less than that of $P(x)$.
The degree of $R(x)$ is less than the degree of $D(x)$.
The degree of $Q(x)$ plus the degree of $D(x)$ equals the degree of $P(x)$.

It happens that these observations hold generally for the division of polynomials, as you can see from the following theorem which we state without proof.

---

**THEOREM**   *(Division Algorithm)*
**5.1**       *Let $P(x)$ be a polynomial and let $D(x)$ be a polynomial with degree smaller than the degree of $P(x)$. Then*

$$P(x) = Q(x)D(x) + R(x)$$

*where D(x) and R(x) are polynomials, and where the de-*
*gree of R(x) is less than that of D(x) and the degree of Q(x)*
*plus the degree of D(x) equals the degree of P(x).*

In searching out zeros of polynomials we will be particularly interested in dividing by first-degree polynomials of the form $x - r$.  Theorem 5.1 tells us that

$$P(x) = Q(x)(x - r) + R(x)$$

and that the degree of R(x) is less than the degree of $x - r$.  But the degree of $x - r$ is 1.  So the degree of R(x) is 0.  That is, R(x) is a constant.
What constant, you ask?  We find out by replacing x by r in P(x).

$$P(r) = Q(r)(r - r) + R(r)$$

So $R(r) = P(r)$.  But R(x) is constant.  So $R(x) = P(r)$ for all x.  We have
just proved the following theorem.

---

**THEOREM
5.2**

$$P(x) = Q(x)(x - r) + P(r)$$

*where P(x) is any nonzero polynomial and where Q(x) is a*
*polynomial of degree one less than the degree of P(x) and*
*where r is any number.*

---

▶ **Illustration 5.15**

Let $P(x) = x^3 + 2x + 3$

$$
\begin{array}{r}
x^2 + 2x + 6 \\
x - 2 \overline{)\ x^3 + 0x^2 + 2x + 3} \\
\underline{x^3 - 2x^2} \\
2x^2 + 2x \\
\underline{2x^2 - 4x} \\
6x + \phantom{0}3 \\
\underline{6x - 12} \\
15
\end{array}
$$

So $x^3 + 2x + 3 = (x^2 + 2x + 6)(x - 2) + 15$.  Plugging 2 into P(x) we get

$$P(2) = 2^3 + 2(2) + 3 = 8 + 4 + 3 = 15$$

Hence $P(r) = R(x)$.  ◀

Theorem 5.3 now follows very easily from Theorem 5.2. It will show you what all of this has to do with zeros of polynomials.

---

**THEOREM 5.3** *If r is a zero of the nonzero polynomial P(x), then*

    **a** *$P(x) = Q(x)(x - r)$, where the degree of Q(x) is one less than the degree of P(x).*

    **b** *All other zeros of P(x) are also zeros of Q(x).*

**PROOF**

We know from Theorem 5.2 that

$$P(x) = Q(x)(x - r) + P(r)$$

with degree of $Q(x)$ one less than that of $P(x)$. But $P(r) = 0$, since $r$ is a zero of $P(x)$. So we get

$$P(x) = Q(x)(x - r)$$

Suppose $s \neq r$ is another zero of $P(x)$:

$$0 = P(s) = Q(s)(s - r)$$

But $s - r \neq 0$. So $Q(s) = 0$. Hence it follows that any remaining zeros of $P(x)$ can be found in $Q(x)$.

---

This theorem arms you with a very useful technique. It says that if you find a zero $r$ of $P(x)$, you can divide $P(x)$ by $x - r$. It will divide evenly (i.e., no remainder), resulting in a quotient $Q(x)$ which has degree one less and which shares any remaining zeros with $P(x)$. Watch how this can be applied.

▶ **Illustration 5.16**

Suppose that you want all the zeros of

$$P(x) = x^3 - 6x^2 + 11x - 6$$

Just by guessing you discover that $P(1) = 0$. So divide $P(x)$ by $x - 1$.

$$
\begin{array}{r}
x^2 - 5x + 6 \\
x - 1 \overline{\smash{)}\, x^3 - 6x^2 + 11x - 6} \\
\underline{x^3 - \phantom{0}x^2} \phantom{+ 11x - 6} \\
-5x^2 + 11x \phantom{- 6} \\
\underline{-5x^2 + \phantom{0}5x} \phantom{- 6} \\
6x - 6 \\
\underline{6x - 6} \\
0
\end{array}
$$

As predicted, the remainder is 0, that is,

$$x^3 - 6x^2 + 11x - 6 = (x^2 - 5x + 6)(x - 1)$$

Theorem 5.3 also predicts that any other zeros of $P(x)$ are zeros of $Q(x) = x^2 - 5x + 6$. We can apply the quadratic formula to $Q(x)$ to find that $x = 2$ and $x = 3$ are the zeros of $Q(x)$. Since any other zeros of $P(x)$ have to be zeros of $Q(x)$, we can conclude that $x = 1$, $x = 2$, and $x = 3$ are the zeros of $P(x)$. ◀

◆ **Illustration 5.17**

Let $P(x) = x^3 - 3x^2 + 1$. Again let us try $x = 1$.

$$
\begin{array}{r}
2x^2 - x - 1 \\
x - 1 \overline{)\, 2x^3 - 3x^2 + 0x + 1} \\
\underline{2x^3 - 2x^2} \\
-\; x^2 + 0x \\
\underline{-\; x^2 + \;\; x} \\
-\; x + 1 \\
\underline{-\; x + 1} \\
0
\end{array}
$$

So we see that 1 is a zero and that

$$P(x) = (x - 1)(2x^2 - x - 1)$$

Any remaining zeros are zeros of $2x^2 - x - 1$ and can be found to be 1 and $-\frac{1}{2}$ (using the quadratic formula). So we see that $-\frac{1}{2}$ is a zero and that 1 is a *repeated* zero ◀

What, you might ask, would one do if he did not guess that $x = 1$ was a zero of the polynomial in Illustration 5.17? The next theorem provides us with a technique for guessing potential zeros—at least rational zeros—of polynomials with coefficients that are integers.

---

**THEOREM 5.4** Let

$$P(x) = a_n x^n + a_{n-1} x^{n-1} + \cdots + a_1 x + a_0$$

*If the rational number $c/d$ (represented in lowest terms) is a zero of $P(x)$, where $c$ and $d$ are integers, then $c$ must be a factor of $a_0$ and $d$ must be a factor of $a_n$.*

---

Since a proof of this theorem does not easily provide insights, we will spend space and time showing you how to use it rather than how to prove it. First, two words of caution.

**1**   The theorem does not guarantee that some rational number is a zero. It only provides a pool of rational numbers which must contain all the rational zeros (if any).

**2**   The theorem does not say anything about the sign of $c/d$. One must test both positive and negative candidates.

**Illustration 5.18**

To find all the rational zeros of $2x^3 - x^2 - 2x + 1$ we need only check numbers of the form $c/d$, where $c$ is a factor of 1 and $d$ is a factor of 2. The factors of 1 are 1 and $-1$, and the factors of 2 are 2, $-2$, 1, and $-1$. So the possible rational zeros of $2x^3 - x^2 - 2x + 1$ are $\frac{1}{1}, \frac{-1}{1}, \frac{1}{2}, \frac{-1}{2}$, that is, $\pm 1$ and $\pm \frac{1}{2}$. Just pick one and try it. (Pick the easy ones first unless your intuition suggests otherwise.)

$$P(1) = 2(1)^3 - (1)^2 - 2(1) + 1 = 0$$

So $x = 1$ is a zero.

Now divide:

$$
\require{enclose}
\begin{array}{r}
2x^2 + x - 1 \\
x - 1 \enclose{longdiv}{2x^3 - x^2 - 2x + 1} \\
\underline{2x^3 - 2x^2} \phantom{-2x+1} \\
x^2 - 2x \phantom{+1} \\
\underline{x^2 - \phantom{2}x} \phantom{+1} \\
-\phantom{2}x + 1 \\
\underline{-\phantom{2}x + 1} \\
0
\end{array}
$$

Any remaining zeros are zeros of $Q(x) = 2x^2 + x - 1$. Applying the quadratic formula to $Q(x)$ we see that

$$x = \frac{-1 + \sqrt{1^2 - (4)(2)(-1)}}{4} = \frac{1}{2}$$

$$= \frac{-1 - \sqrt{1^2 - (4)(2)(4)}}{4} = -1$$

are zeros. Hence 1, $\frac{1}{2}$, and $-1$ are the zeros of $P(x)$.

**Illustration 5.19**

Now let us try $P(x) = 3x^4 - 7x^3 + 5x^2 - 7x + 2$. The factors of 2 are $\pm 2$ and $\pm 1$, and the factors of 3 are $\pm 3$ and $\pm 1$. So possible rational zeros are $\pm \frac{2}{3}, \pm \frac{1}{3}, \pm 2$, and $\pm 1$.

Checking you will find that $P(1) \neq 0$ and $P(-1) \neq 0$. But $P(2) = 0$.

Dividing

$$
\begin{array}{r}
3x^3 - x^2 + 3x - 1 \\
x - 2 \overline{)\, 3x^4 - 7x^3 + 5x^2 - 7x + 2} \\
\underline{3x^4 - 6x^3} \\
-\ x^3 + 5x^2 \\
\underline{-\ x^3 + 2x^2} \\
3x^2 - 7x \\
\underline{3x^2 - 6x} \\
-x + 2 \\
\underline{-x + 2} \\
0
\end{array}
$$

Looking at $Q(x) = 3x^3 - x^2 + 3x - 1$ you see that the possible rational zeros are $\pm 1$ and $\pm\frac{1}{3}$. We already know that $Q(1) \neq 0$ and $Q(-1) \neq 0$ [otherwise $P(1)$ or $(-1)$ would have been 0]. So, we try $\frac{1}{3}$ and find that $Q(\frac{1}{3}) = 0$.

Dividing,

$$
\begin{array}{r}
3x^2 + 3 \\
x - \tfrac{1}{3} \overline{)\, 3x^3 - x^2 + 3x - 1} \\
\underline{3x^3 - x^2} \\
0 + 3x - 1 \\
\underline{3x - 1} \\
0
\end{array}
$$

So, we know that any additional zeros must be zeros of $S(x) = 3x^2 + 3$ which you can see has no real zeros ($b^2 - 4ac < 0$). We can conclude that the only rational zeros of $P(x)$ are $x = 2$ and $x = \frac{1}{3}$. Moreover, these are the only real zeros, since $S(x)$ has no real zeros. ◄

◄ **Illustration 5.20**

The possible rational zeros of $x^4 - 5x^2 + 6$ are $\pm 6, \pm 3, \pm 2,$ and $\pm 1$, none of which works. What do you do? Either you are clever and notice that $x^4 - 5x^2 + 6 = (x^2 - 2)(x^2 - 3)$ so that it has $\pm\sqrt{2}$ and $\pm\sqrt{3}$ as zeros, or you apply the techniques of the next sections. ◄

Summarizing this section we have found that

If $r$ is a zero of the polynomial $P(x)$, then $P(x) = (x - r)\,Q(x)$, where $Q(x)$ has degree one less than the degree of $P(x)$.

Furthermore, any rational roots of $P(x) = a_n x^n + a_{n-1} x^{n-1} + \cdots + a_1 x + a_0$ where $a_0, a_1, \ldots, a_{n-1}, a_n$ are integers can be found among the numbers $c/d$, where $c$ is a factor of $a_0$ and $d$ is a factor of $a_n$.

So that in our sleuthing for zeros of $P(x)$ we can

Look (by trial and error) for any rational zeros from among the numbers $c/d$ described above.

Whenever a zero ($r$) is found, divide $P(x)$ by $x - r$.

Repeat the above process on the quotient $Q(x)$ until, perhaps, $P(x)$ is reduced to a second-degree polynomial.

Apply the quadratic formula to the second-degree polynomial.

What do you do when you run out of rational zeros? You will see in Sec. 5.7.

## PROBLEMS 5.4

**1** Use Theorem 5.2 and division to find each of the following.
  **a** Let $P(x) = x^4 - 3x^3 + 2x^2 + 1$. Find $P(4)$.
  **b** Let $P(x) = 4x^3 - 2x^2 - 5$. Find $P(13)$.
**2** Use Theorem 5.2 and division to find each of the following.
  **a** Let $P(x) = 13x^5 - 4x^4 - 2x^3 + 3x - 2$. Find $P(7)$.
  **b** Let $P(x) = x^4 - 2$. Find $P(11)$.
**3** Use all the techniques at your disposal to find as many zeros as possible of the following polynomials.
  **a** $P(x) = x^2 + 3x + 1$
  **b** $P(x) = x^3 + 2x^2 - 2x - 1$
  **c** $P(x) = 2x^3 - 3x^2 - 2x + 6$
  **d** $P(x) = 2x^4 - x^3 - 5x^2 + 4x + 6$
  **e** $P(x) = x^4 + 3x^3 - x^2 - 6x - 2$
**4** Use all the techniques at your disposal to find as many zeros as possible of the following polynomials.
  **a** $P(x) = 2x^2 - 6$
  **b** $P(x) = 2x^3 - 6x$
  **c** $P(x) = 4x^3 - 10x^2 - 12x + 30$
  **d** $P(x) = 2x^4 + 3x^3 - 3x^2 - 3x + 2$
  **e** $P(x) = 2x^4 - 13x^2 + 15$
**5** Describe and justify each step involved in expressing $P(x) = x^3 + 6x^2 + 11x + 6$ in the form $P(x) = (x - r_1)(x - r_2)(x - r_3)$.
**6** Describe and justify each step involved in expressing $P(x) = x^3 - 4x^2 - 7x + 10$ in the form $P(x) = (x - r_1)(x - r_2)(x - r_3)$.

## 5.5    FACTORING POLYNOMIALS

You already have some experience factoring whole numbers. For example, $24 = 4 \cdot 6 = 2 \cdot 2 \cdot 2 \cdot 3$. In general, you are expressing the whole number as a product of simpler whole numbers. In Sec. 5.4 you saw that if $r$ is a zero of the polynomial, then

$$P(x) = (x - r)Q(x)$$

This is also a form of factoring and is very closely related to the process of finding the zeros of a polynomial. This section will give you an opportunity to develop some skill with factoring, as well as reinforce the concepts and skills of Sec. 5.4.

### Factoring Polynomials of Degree 1

If $P(x) = ax + b$, $P(x)$ can be expressed in the form $P(x) = a[x - (-b/a)]$, which is its factored form.

◆ **Illustration 5.21**

If $P(x) = -3x + 4$, $P(x) = -3(x - \frac{4}{3})$
Note that $\frac{4}{3}$ is a zero of $P(x)$. ◀

### Factoring Polynomials of Degree 2

You already know that

$$r_1 = \frac{-b + \sqrt{b^2 - 4ac}}{2a} \quad \text{and} \quad r_2 = \frac{-b - \sqrt{b^2 - 4ac}}{2a}$$

are the zeros of the polynomial $ax^2 + bx + c$. You also know that, if $r$ is a zero of a polynomial, $x - r$ is a factor. So you can conclude that $x - r_1$ and $x - r_2$ are both factors of $ax^2 + bx + c$.

Multiplying, we get

$$(x - r_1)(x - r_2) = \left(x - \frac{-b + \sqrt{b^2 - 4ac}}{2a}\right)\left(x - \frac{-b - \sqrt{b^2 - 4ac}}{2a}\right)$$

$$= x^2 + \frac{b}{a}x + \frac{c}{a}$$

Multiplying through by $a$, we get

$$ax^2 + bx + c = a\left(x - \frac{-b + \sqrt{b^2 - 4ac}}{2a}\right)\left(x + \frac{-b - \sqrt{b^2 - 4ac}}{2a}\right)$$

This gives us a general formula for factoring any second-degree polynomial.

◆ **Illustration 5.22**

If $P(x) = 3x^2 + 4x + 1$,

$$P(x) = 3\left(x - \frac{-4 + \sqrt{16 - 12}}{6}\right)\left(x - \frac{-4 - \sqrt{16 - 12}}{6}\right)$$

$$= 3[x - (-\tfrac{1}{3})][x - (-1)]$$

Note again that $-\frac{1}{3}$ and $-1$ are zeros of $P(x)$. ◀

In the method illustrated here second-degree polynomials $P(x)$ are factored by finding their zeros ($r_1$ and $r_2$) using the quadratic formula and then expressing the polynomial as $P(x) = a(x - r_1)(x - r_2)$, where $a$ is the coefficient of $x^2$.

Sometimes this process can be reversed, and factoring can be used to discover the zeros of a quadratic polynomial. You will be able to factor certain quadratics by inspection, once you have gained some experience.

We have learned how to multiply

$$(x - 3)(x - 2) = x^2 - 5x + 6$$

We want to be able to reverse the process, that is, to start with $x^2 - 5x + 6$ and end up with $(x - 3)(x - 2)$. To learn this factoring process we analyze the multiplication process.

$$(ax + b)(cx + d) = acx^2 + (ad + cb)x + bd$$

### Finding Coefficients

**1** To find the coefficients of $x$($a$ and $c$) we look to the factors of the $x^2$ coefficient ($ac$).

**2** To find the constant coefficients ($b$ and $d$) we look to the factors of the constant coefficient ($bd$).

**3** To determine which coefficients go together and with what sign we use trial and error and check our result. If this procedure does not work, we can fall back on the method described above in factoring polynomials of degree 2 that uses the quadratic formula.

◆ **Illustration 5.23**

$$x^2 - 3x + 2 = (ax + b)(cx + d)$$

$a$ and $c$ have to be factors of the $x^2$ coefficient which is 1; try 1 for each.

$b$ and $d$ have to be factors of 2, so try 2 for one and 1 for the other.

To this point we have

$$x^2 - 3x + 2 = [x + (\pm 2)][x + (\pm 1)]$$

The $-3$ and the $+2$ (in $x^2 - 3x + 2$) tip us off that we need both of the constant coefficients to be negative. In fact you will see that $x^2 - 3x + 2 = (x - 2)(x - 1)$ ◀

◆ **Illustration 5.24**

Factor $x^2 + 6x + 8$.

$$x^2 + 6x + 8 = (ax + b)(cx + d)$$

The best guess for $a$ and $c$ is 1. The product of $b$ and $d$ must be 8. Candidates are $(\pm 1, \pm 8)$, $(\pm 2, \pm 4)$.

$$(x + b)(x + d) = x^2 + (b + d)x + bd$$

so we need $bd = 8$ and $b + d = 6$. A reasonable guess is $b = 2$ and $d = 4$. Indeed, $(x + 2)(x + 4) = x^2 + 6x + 8$. ◀

### ▶ Illustration 5.25

$$8x^2 + 2x - 15 = (ax + b)(cx + d)$$

We need $ac = 8$, so the pairs $(\pm 1, \pm 8)$, $(\pm 2, \pm 4)$ are candidates. We also need $bd = -15$, so the pairs $(\pm 1, \pm 15)$, $(\pm 3, \pm 5)$ are candidates.

Then we need to choose signs and pairs so that $ad + bc = 2$. If you guess $(2x - 3)(4x + 5)$, you are close.

$$(2x - 3)(4x + 5) = 8x^2 - 2x - 15$$

Switch the signs and you get $(2x + 3)(4x - 5) = 8x^2 + 2x - 15$. To put this in the form $(x - r_1)(x - r_2)$ we write

$$(2x + 3) = 2[x - (-\tfrac{3}{2})] \qquad \text{and} \qquad (4x - 5) = 4(x - \tfrac{5}{4})$$

so that $8x^2 + 2x - 15 = 8[x - (-\tfrac{3}{2})](x - \tfrac{5}{4})$ ◀

### ▶ Illustration 5.26

Try

$$x^2 - 4 = (x - 2)(x + 2) \quad ◀$$

### ▶ Illustration 5.27

Try

$$6x^2 - 2x - 28 = (2x + 4)(3x - 7) = 6[x - (-2)](x - \tfrac{7}{3}) \quad ◀$$

Experience is a very good teacher for these factoring problems.

### Factoring Polynomials of Degree Higher Than 2

Here we need to rely in general on our ability to locate the zeros of a polynomial using the methods of Sec. 5.4. We know that, if $r_1$ is a zero of $P(x)$, $P(x) = (x - r_1)Q_1(x)$, where the remaining zeros of $P(x)$ are zeros of $Q_1(x)$. Then if $r_2$ is a zero of $Q_1(x)$, $Q_1(x) = (x - r_2)Q_2(x)$. Putting the last two sentences together we get

$$P(x) = (x - r_1)Q_1(x)$$
$$= (x - r_1)(x - r_2)Q_2(x)$$

Each time we find another zero, we can reduce the degree of the polynomial $Q(x)$ by one. If the degree of $Q(x)$ is reduced to 2, the quadratic formula or factoring can be used to factor $Q(x)$. All this leads up to a theorem which we will state without proof.

---

**THEOREM 5.5**  *If $P(x)$ is a polynomial of degree n, where $n \geq 1$, then $P(x) = a(x - r_1)(x - r_2) \cdots (x - r_n)$, and $r_1, r_2, \ldots, r_n$ are the n zeros of $P(x)$. (It is important to note that some of the r's may be repeated and some may be complex numbers.)*

---

▶ **Illustration 5.28**

Let $P(x) = 4x^4 + 12x^3 + 7x^2 - 3x - 2$. Possible rational zeros include $\pm 1, \pm 2, \pm \frac{1}{2}, \pm \frac{1}{4}$.

Checking, we find that $\frac{1}{2}$ is a zero.

$$
\begin{array}{r}
4x^3 + 14x^2 + 14x + 4 \\
x - \tfrac{1}{2} \overline{)\ 4x^4 + 12x^3 + 7x^2 - 3x - 2} \\
\underline{4x^4 - \phantom{1}2x^3} \\
14x^3 + 7x^2 \\
\underline{14x^3 - 7x^2} \\
14x^2 - 3x \\
\underline{14x^2 - 7x} \\
4x - 2 \\
\underline{4x - 2} \\
0
\end{array}
$$

So $P(x) = (x - \frac{1}{2})(4x^3 + 14x^2 + 14x + 4)$.

Checking again we find that $-2$ is a zero of $4x^3 + 14x^2 + 14x + 4$. So dividing by $x - (-2) = x + 2$ we get

$$\frac{4x^3 + 14x^2 + 14x + 4}{x + 2} = 4x^2 + 6x + 2$$

Now we have

$$P(x) = (x - \tfrac{1}{2})(4x^3 + 14x^2 + 14x + 4)$$
$$= (x - \tfrac{1}{2})[x - (-2)](4x^2 + 6x + 2)$$

Now, we can factor

$$4x^2 + 6x + 2 = 2(2x^2 + 3x + 1)$$
$$= 2(2x + 1)(x + 1)$$
$$= 4[x - (-\tfrac{1}{2})][x - (-1)]$$

So we have

$$P(x) = 4(x - \tfrac{1}{2})[x - (-2)][x - (-\tfrac{1}{2})][x - (-1)]$$

where we have rewritten each factor in the form $x - r$. (Note that $2x + 1 = 2(x + \tfrac{1}{2}) = 2[x - (-\tfrac{1}{2})].$) ◀

## PROBLEMS 5.5

**1** Factor the following polynomials of degree 2.
  **a** $P(x) = x^2 - 13x + 36$
  **b** $P(x) = 3x^2 + 5x + 2$
  **c** $P(x) = x^2 + 7x + 11$
  **d** $P(x) = 9x^2 - 16$
  **e** $P(x) = 3x^2 - 5x + 1$
**2** Factor the following polynomials of degree 2.
  **a** $P(x) = x^2 - 3x - 4$
  **b** $P(x) = 6x^2 - 13x - 5$
  **c** $P(x) = x^2 + x + 1$
  **d** $P(x) = 4x^2 - 9$
  **e** $P(x) = 2x^2 - 12x - 4$
**3** Factor the polynomial $x^3 + x^2 - 5x + 3$.
**4** Factor the polynomial $x^3 - 3x^2 + 4$.

## 5.6   GRAPHING POLYNOMIAL FUNCTIONS

The graphing of polynomial functions will involve plotting some points, i.e., finding $P(r)$ for certain values of $r$ so that $(r, P(r))$ can be placed on the graph. In particular we will be able to take advantage of what we know about the zeros of a polynomial function [the points $(x, 0)$ where its graph crosses, or touches, the $X$ axis] to help determine the shape of the graph. Then in Sec. 5.7 we will be able to turn around and take advantage of what we know about the graph of a polynomial to help approximate its zeros—especially its irrational zeros.

First we would like to make some general and informal statements about the graphs of polynomials.

The graph of a polynomial is smooth. It may have bumps in it, but it will not have any sharp edges or breaks. Figure 5.1 roughly shows the kinds of shapes polynomial graphs can have.

Between two zeros of a polynomial there will be at least one bump. However, you may have a bump and yet have no zero and you may have a single zero with no bump. Figure 5.2 gives an example of each of these phenomena.

The term with the largest power of $x$ dominates a polynomial for large values of $|x|$. So, for example, $P(x) = 5x^3 - 3x^2 + 2x - 7$ will get

**Figure 5.1**

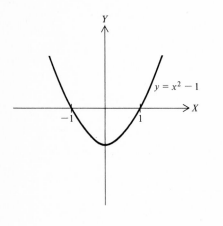

$y = x^2 - 1$

$-1$   $1$

Two zeros with a bump between

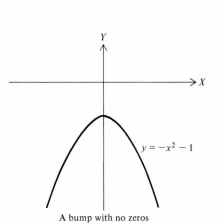

$y = -x^2 - 1$

A bump with no zeros

$y = x^3$

A zero but no bump

**Figure 5.2**

very large as x gets large and will get negatively large as x gets negatively large.  See Fig. 5.3.  This is because the 5x³ term behaves that
way.  (Note the odd exponent of x.)

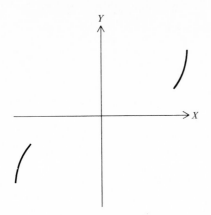

P(x) gets larger for larger x and
negatively large for negatively large x

**Figure 5.3**

On the other hand, $P(x) = x^4 + 5x^3 - 7x^2 + 16$ will get large as x gets
large or negatively large because x⁴ behaves that way.  (Note the even
exponent of x.)  See Fig. 5.4.

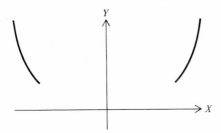

P(x) gets large as x gets large and as
x gets negatively large

**Figure 5.4**

In the light of the comments made here we suggest the following
procedure for graphing a polynomial P(x).

### Graphing a Polynomial

**1**   Find the zeros of P(x).

**2**  Analyze the behavior of $P(x)$ for $x$ large and for $x$ negatively large.

**3**  Compute $P(x)$ and plot $(x, P(x))$ where you need to in order to complete your understanding of the shape of the graph.†

**4**  Sketch the graph.

◆ **Illustration 5.29**

Let $P(x) = x^3 - 6x^2 + 11x - 6$.

You will find that the zeros of $P(x)$ are 1, 2, and 3; that is, $(1, 0)$, $(2, 0)$, and $(3, 0)$ lie on the graph of $y = P(x)$.

For $x$ large $P(x)$ will get large, and for $x$ negatively large $P(x)$ will get negatively large.  So we roughly know what is shown in Fig. 5.5.

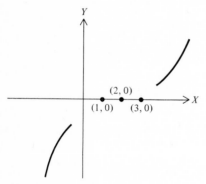

**Figure 5.5**

You also know that there have to be bumps between the zeros.  It makes sense to check out a point on the left of 1, for example, $x = 0$; a point between 1 and 2, for example, $x = \frac{3}{2}$; a point between 2 and 3, for example, $x = \frac{5}{2}$; and a point to the right of 3, for example, $x = 4$.  You may want to verify the values in Table 5.1 using a hand calculator, synthetic division (Appendix A), or a lot of paper-and-pencil arithmetic.

**Table 5.1**

| $x$ | $-1$ | 0 | $\frac{3}{2}$ | $\frac{5}{2}$ | 4 |
|---|---|---|---|---|---|
| $P(x)$ | $-24$ | $-6$ | $\frac{3}{8}$ | $-\frac{3}{8}$ | 6 |

Plotting these points on the graph together with our zeros enables us to sketch Fig. 5.6 fairly confidently.    ◆

---

† If you have a hand calculator, it will come in handy for computing $P(x)$'s.  If not, you may want to learn the technique of synthetic division in Appendix A.

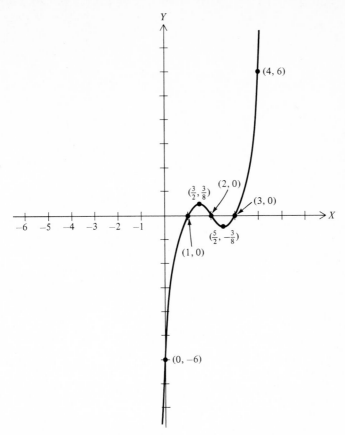

**Figure 5.6**

◆ **Illustration 5.30**

Let us graph $P(x) = x^4 + 4x^3 - 2x^2 - 8x$. Noticing that $P(x) = x(x^3 + 4x^2 - 2x - 8)$, we find that $x = 0$ is a zero and then we find that $x = -4$, $x = -\sqrt{2}$, and $x = \sqrt{2}$ are also zeros.

Since $x^4$ is an even power of $x$, it gets large when $x$ gets large and when $x$ gets negatively large. So $P(x)$ will do the same. We now know the rough information in Fig. 5.7. Let us find $(-5, P(-5))$, $(-2, P(-2))$, $(-1, P(-1))$, $(1, P(1))$, and $(2, P(2))$. (See Table 5.2.) Adding these points to our graph we sketch Fig. 5.8.

**Table 5.2**

| x | −5 | −3 | −2 | −1 | 1 | 2 |
|------|-----|-----|-----|-----|-----|-----|
| P(x) | 115 | −21 | −4 | 3 | −5 | 24 |

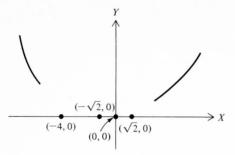

**Figure 5.7**

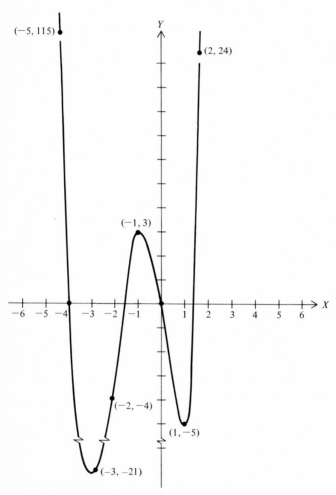

**Figure 5.8**

If in any of your graphing work you desire more detail and accuracy, you can just plot additional points. ◀

The following problems give you a chance to develop your graphing skills before they are utilized in the next section in the process of approximating elusive zeros of polynomials.

## PROBLEMS 5.6

1 Review your graphing skills by graphing the following simple polynomials.

    **a** $P(x) = -2$    **b** $P(x) = 2x - 3$    **c** $P(x) = -x^2 + 3x - 2$

2 Graph the following polynomials of degree 0, 1, and 2.

    **a** $P(x) = \frac{5}{2}$    **b** $P(x) = 4 - 3x$    **c** $P(x) = x^2 + 2x + 2$

3 Graph $P(x) = x$, $P(x) = x^2$, $P(x) = x^3$, and $P(x) = x^4$. Then try to generalize about the graph of $P(x) = x^n$ for any positive integer $n$.

4 Graph the following polynomials using any techniques that you know.

    **a** $P(x) = x^3 - 2x^2 - 5x + 6$
    **b** $P(x) = 2x^3 + x^2 - 10x - 5$
    **c** $P(x) = 2x^4 - 4x^3 - 10x + 12x$

5 Graph the following polynomials.

    **a** $P(x) = x^3 + 2x^2 - 5x - 6$
    **b** $P(x) = 4x^3 + 6x^2 - 6x - 9$
    **c** $P(x) = 3x^4 + 6x^3 - 15x^2 - 18x$

6 Graph the following, somewhat trickier, polynomials.

    **a** $P(x) = x^3 - 3x^2 + 3x - 1$
    **b** $P(x) = x^4 + 2x^2 + 1$
    **c** $P(x) = x^5 - 1$

## 5.7    APPROXIMATING ZEROS

As you have seen in Sec. 5.6, the graphs of polynomials are smooth curves without any breaks in them. So for a polynomial to change from positive to negative it must go through a zero. See Fig. 5.9, for example.

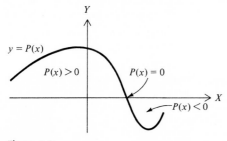

**Figure 5.9**

We are going to use this fact to help us in approximating zeros of polynomials. You may ask, "Why approximate zeros when we already know how to find them?" The answer is that we have learned how to find rational numbers that are zeros of polynomials. We have even learned how to find irrational zeros of polynomials that have enough rational roots to enable us to reduce the polynomials to second degree. Some polynomials, however, do not have enough rational roots, or any rational roots for that matter. We will present the material of this section by means of illustrations.

**Illustration 5.31**

Let $P(x) = x^2 - 2$. We begin with this simple polynomial for which we know the zeros so that it will be easy to see what is happening. We know that $P(x)$ has zeros at $\sqrt{2}$ and $-\sqrt{2}$ and that its graph looks like Fig. 5.10.

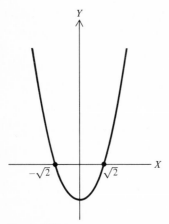

Figure 5.10

Suppose we did not know so much. First we would find points where $P(x)$ has opposite signs; for example, $P(1) = -1$ and $P(2) = 2$. We know that there must be a zero of $P(x)$ between 1 and 2. Try 1.5. $P(1.5) = 0.25$. Since $P(1) = -1$ and $P(1.5) = 0.25$, there must be a zero between 1 and 1.5. (See Fig. 5.11.) Moreover, since $P(1.5)$ is closer to zero than $P(1)$ is, let us guess that the zero is closer to 1.5 than it is to 1. Let us try 1.4. $P(1.4) = -0.04$. Now since $P(1.4) = -0.04$ and $P(1.5) = 0.25$, we know that there is a zero between 1.4 and 1.5. Moreover, it is likely to be closer to 1.4 than to 1.5, and so on. You know that the furthest you are from a zero is the distance between 1.4 and 1.5, which is

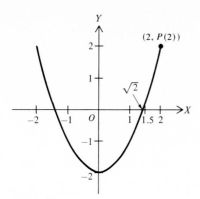

**Figure 5.11**

0.1. If you need to get closer, try 1.42. $P(1.42) = 0.0164$. We now know that there is a zero between 1.4 and 1.42. So if you choose 1.41 as an approximation to a zero, you cannot be off by more than $1.42 - 1.4 = 0.02$. We have found a fair approximation to $\sqrt{2}$. ◀

### ▶ Illustration 5.32

Now let us try one where we do not know what is going on in advance. Let $P(x) = x^4 + 3x^3 - x^2 - 6x - 2$. The possible rational zeros are $\pm 1$ and $\pm 2$. You check and you find that none of these is a zero, so you are at a loss for more candidates. You do find, however, that $P(-2) = -2$ and $P(-3) = 7$, so that there must be a zero (irrational) between $-2$ and $-3$. It is probably closer to $-2$. (See Fig. 5.12.)

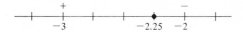

**Figure 5.12**

Try $-2.25$. We find using a hand calculator, synthetic division, or perspiration, that $P(-2.25) < 0$. So you know that you have a zero between $-3$ and $-2.25$. Try $-2.5$. $P(-2.5) < 0$. So our guess that the zero was closer to $-2$ than to $-3$ was in error. Try $-2.75$. Finally, $P(-2.75) > 0$.

The search for a closer approximation to a zero can now be carried on between $-2.75$ and $-2.5$. This is one of the exercises at the end of this section. ◀

In summary, to find the zeros of a polynomial you

First check all possible rational zeros and divide by $x - r$, if r is a zero.

If you can reduce the polynomial to a second-degree polynomial using rational zeros, then apply the quadratic formula to get the remaining zeros.

If you cannot reduce the polynomial to a second-degree polynomial, then the remaining real zeros can be approximated using the techniques of this section.

## PROBLEMS 5.7

**1** Continue the approximation started in Illustration 5.32 to get within 0.05 of a zero of $P(x)$.

**2** Approximate a different zero of $P(x)$ in Illustration 5.32.

**3** Approximate a zero of $P(x) = 2x^3 - x^2 - 10x + 5$, which is between 2 and 3 to within 0.05. Is your approximation larger or smaller than the zero?

**4** Approximate a zero of $P(x) = x^4 - 10x^2 + 21$, which lies between $-1$ and $-2$ to within 0.05. Is your approximation larger or smaller than the zero?

# EXPONENTIAL FUNCTIONS

## 6.1    A SIMPLE EXPONENTIAL FUNCTION

You have already been introduced to a broad class of functions, namely the polynomial functions, of which $x^2$ is an example. In this chapter you will meet with another important class of functions called the *exponential* functions, of which $2^x$ is an example. The name comes from the fact that the variable is an exponent. We shall discuss several exponential functions, graphing them and indicating their properties and applications.

▶ **Illustration 6.1**

Let us begin with the simple function $f$ such that $y = f(x) = 2^x$. We first compute some values as given in Table 6.1. (Review Secs. 1.10 and 1.11.)

**Table 6.1**

| x | -3 | -2 | -1 | $-\frac{1}{2}$ | $-\frac{1}{3}$ | 0 | $\frac{1}{3}$ | $\frac{1}{2}$ | 1 | 2 | 3 |
|---|---|---|---|---|---|---|---|---|---|---|---|
| $2^x$ | $\frac{1}{8}$ | $\frac{1}{4}$ | $\frac{1}{2}$ | 0.707 | 0.794 | 1 | 1.260 | 1.414 | 2 | 4 | 8 |

Here are the details of the computations.

$$2^{-3} = \frac{1}{2^3} = \frac{1}{8} \qquad 2^3 = 8$$

$$2^{-2} = \frac{1}{2^2} = \frac{1}{4} \qquad 2^2 = 4$$

$$2^{-1} = \frac{1}{2^1} = \frac{1}{2} \qquad 2^0 = 1$$

$$2^{-1/2} = 1/2^{1/2}$$

To simplify this we multiply both numerator and denominator by $2^{1/2}$, getting

$$2^{-1/2} = \frac{2^{1/2}}{2^{1/2} \times 2^{1/2}}$$

where the denominator is now $2^{1/2+1/2} = 2^{2/2} = 2$.

So
$$2^{-1/2} = \frac{2^{1/2}}{2} = \frac{1}{2}\sqrt{2}$$
$$= \tfrac{1}{2}(1.414) \qquad \text{Table VII, Appendix B}$$
$$= 0.707$$
$$2^{-1/3} = 1/2^{1/3}$$

To simplify this we multiply both numerator and denominator by $2^{2/3}$, getting

$$2^{-1/3} = \frac{2^{2/3}}{2^{1/3} \times 2^{2/3}}$$

where the denominator is now $2^{1/3+2/3} = 2^{3/3} = 2$.

So
$$2^{-1/3} = \tfrac{1}{2}(2^{2/3}) = \tfrac{1}{2}\sqrt[3]{2^2}$$
$$= \tfrac{1}{2}\sqrt[3]{4} = \tfrac{1}{2}(1.587) \qquad \text{Table VII, Appendix B}$$
$$= 0.794$$
$$2^{1/3} = \sqrt[3]{2}$$
$$= 1.260$$
$$2^{1/2} = \sqrt{2}$$
$$= 1.414$$

The points of Table 6.1 are plotted in Fig. 6.1, and the curve is

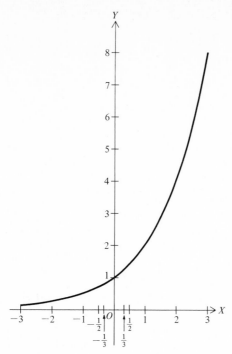

**Figure 6.1**

sketched.   The graph lies wholly above the X axis: $2^x$ is positive for any real number x, but we do not prove it here.   ◀

## PROBLEMS 6.1

*In Probs. 1 to 4 fill in the table of values of the indicated functions.*

**1**

| x | $-2$ | $-1$ | $-\frac{1}{2}$ | $-\frac{1}{3}$ | 0 | $\frac{1}{3}$ | $\frac{1}{2}$ | 1 | 2 |
|---|------|------|----------------|----------------|---|---------------|---------------|---|---|
| $x^2$ | | | | | | | | | |

**2**

| x | $-2$ | $-1$ | $-\frac{1}{2}$ | $-\frac{1}{3}$ | 0 | | $\frac{1}{2}$ | 1 | 2 |
|---|------|------|----------------|----------------|---|---|---------------|---|---|
| $3^x$ | | | | | | | | | |

**3**

| x | $-2$ | $-1$ | $-\frac{1}{2}$ | $-\frac{1}{3}$ | 0 | $\frac{1}{3}$ | $\frac{1}{2}$ | 1 | 2 |
|---|------|------|----------------|----------------|---|---------------|---------------|---|---|
| $2^{-x}$ | | | | | | | | | |

**4**

| x | $-2$ | $-1$ | $-\frac{1}{2}$ | $-\frac{1}{3}$ | 0 | $\frac{1}{3}$ | $\frac{1}{2}$ | 1 | 2 |
|---|------|------|----------------|----------------|---|---------------|---------------|---|---|
| $x^3$ | | | | | | | | | |

*In Probs. 5 to 8 sketch the two curves on the same axes and to the same scales using the values of x indicated in Probs. 1 to 4 above.*

**5**   **a**   $y = 2^x$          **6**   **a**   $y = 3^x$

      **b**   $y = x^2$                **b**   $y = x^3$

**7   a**   $y = 2^x$      **8   a**   $y = 2^x$

    **b**   $y = 3^x$          **b**   $y = 2^{-x}$

## 6.2    EXPONENTIAL FUNCTIONS

In the previous section we spent some time with the function defined by $y = 2^x$. This is a special case of the general exponential function defined as follows.

### DEFINITION 6.1

---

The function defined by $y = a^x (a > 0)$ is called the *exponential function with base a*. Its domain is the set of real numbers, and for $a \neq 1$ its range is the set of positive reals. A more general form of the exponential function is given by $y = ka^{bx}$, where $k \neq 0$ and $b \neq 0$ are constants. (See Fig. 6.1 for $y = 2^x$.) ◀

From Sec. 1.10 we know the meaning of $a^x$ for $a > 0$ and x a rational number. Recall that $a^{p/q}$ is defined to be $a^{p(1/q)} = (a^p)^{1/q}$, where p and q are integers. To extend the theory to include $a^x$ for x an irrational, and therefore for any real number x, is beyond the scope of this book. However, a natural way to study a number such as $2^\pi$, for example, would be to consider the successive decimal approximations to $\pi$, such as 3.1, 3.14, 3.141, 3.1415, etc. Then $2^{3.1}$, $2^{3.14}$, $2^{3.141}$, $2^{3.1415}$, etc. are successive approximations to $2^\pi$.

◆ **Illustration 6.2**

The following computations were done on a hand calculator. (The early entries could be done by using a standard five-place table of logarithms but not with the same accuracy.)

$$2^{3.1} \quad\quad\quad = 8.57418770$$
$$2^{3.14} \quad\quad\; = 8.81524093$$
$$2^{3.141} \quad\quad = 8.82135330$$
$$2^{3.1415} \quad\; = 8.82441109$$
$$2^{3.14159} \quad = 8.82496160$$
$$2^{3.141592653} = 8.82497782$$

The last four approximations to $2^\pi$ all begin with 8.82 which is possibly a reasonable approximation to $2^\pi$ for most purposes. ◀

The following operations hold and are similar to those in Sec. 1.10. For all real numbers $x$ and $y$:

---

**THEOREM 6.1**  $\qquad a^x a^y = a^{x+y} \qquad \dfrac{a^x}{a^y} = a^{x-y}$

**THEOREM 6.2**  $\qquad a^{-x} = \dfrac{1}{a^x} \qquad a^0 = 1$

**THEOREM 6.3**  $\qquad a^{m/n} = \sqrt[n]{a^m} = (\sqrt[n]{a})^m \qquad m, n$ positive integers

**THEOREM 6.4**  $\qquad (a^x)^y = a^{xy} \qquad (ab)^x = a^x b^x \qquad b$ positive

---

◆ **Illustration 6.3**

Here is an illustration of how these properties may be used to effect a simplification. Examine each step carefully.

$$2^x \cdot 4^{2x} = 2^x(2^2)^{2x} = 2^x(2^{4x}) = 2^{5x} = (2^5)^x = 32^x \quad ◀$$

**PROBLEMS 6.2**

In Probs. 1 to 4 simplify but leave your answer in the form $a^x$.

**1** $2^4 \cdot 4^{-2} \cdot \sqrt{2}$  **2** $3^2 \cdot 9^{-4} \cdot 27^2$  **3** $8^2 \cdot 8^{1/2} \cdot 64$  **4** $25/5^3$

In Probs. 5 to 20 simplify where possible.

**5** $2(2^x)$  **6** $7^x \cdot 7^{2x}$  **7** $3^x/3^{-2x}$

**8** $4^x + 4^{2x}$  **9** $8(2^x)(2^{2x})$  **10** $3^{3x} \cdot 9^{2x} \cdot 27^{3x}$

**11** $(x^2)(2^x)$  **12** $xa^{-x}$  **13** $3a^{2x+5}$

**14** $-2a^{3-x}$  **15** $a^{2x} \cdot a^{-3x} \cdot a^{x/2}$  **16** $a^{3x} \cdot a^{-4x} \cdot a^{2x}$

**17** $b^{x/2} \cdot b^{x/3}$  **18** $b^{-x}(b^x)^2$  **19** $(c^{1/2})^x \cdot c^{-2x}$

**20** $c^{-x} \cdot c^{-2x} \cdot c^{7x}$

**6.3  A VERY SPECIAL EXPONENTIAL FUNCTION**

By all odds the most important exponential function is the function defined by $y = e^x$, where $e$ is a certain irrational number, namely, correct to five decimal places, $e = 2.71828$. It ranks, along with $\pi$, as one of the most important constants in mathematics. Table II, Appendix B, gives values of $e^x$ and $e^{-x}$, and Table IV, Appendix B, lists the so-called *natural* logarithms with base $e$ instead of base 10. It is impossible to explain why the number $e$ and the associated exponential function defined by $y = e^x$ are of such importance. In fact it is so important that we speak of it as being *the* exponential function. A more general form of this exponential function is given by $y = ke^{bx}$, where $k \neq 0$ and $b \neq 0$ are constants.

A convenient approximation to $e^x$ is afforded by the polynomial

$$1 + \frac{x}{1!} + \frac{x^2}{2!} + \frac{x^3}{3!} + \cdots + \frac{x^n}{n!}$$

where $1! = 1$, $2! = 1 \cdot 2$, $3! = 1 \cdot 2 \cdot 3$, and $n! = 1 \cdot 2 \cdot 3 \cdots n$, and where $n!$ is read "$n$ factorial." As $n$ increases the approximation becomes closer and closer. In fact $e^x$ is given by an infinite sum:

$$e^x = 1 + \frac{x}{1!} + \frac{x^2}{2!} + \frac{x^3}{3!} + \cdots + \frac{x^n}{n!} + \cdots \tag{1}$$

which cannot be explained in this book.

It is rather a simple matter to sketch the graph of an exponential function over a small interval about the origin. We illustrate by sketching both $y = e^x$ and $y = e^{-x}$ in the interval $-3 \leq x \leq 3$.

From Table II, Appendix B, we prepare a smaller table first for $e^x$ as below. Note that when $x = 2$ we find $e^2$ in the $e^x$ column and the value of $e^{-2}$ in the $e^{-x}$ column.

Our smaller table looks like Table 6.2 (up to the double vertical line).

The numbers in the $e^x$ column are the same numbers as in the $e^{-x}$ column — but they are reversed. The graphs are indicated in Fig. 6.2.

**Table 6.2**

| $x$ | $e^x$ | $y = e^x$ (rounded) | $y = e^{-x}$ (rounded) |
|------|---------|------|------|
| 2 | 7.3891 | 7.4 | 0.1 |
| 1.5 | 4.4817 | 4.5 | 0.2 |
| 1 | 2.7183 | 2.7 | 0.4 |
| 0.5 | 1.6487 | 1.6 | 0.7 |
| 0 | 1.0000 | 1.0 | 1.0 |
| −0.5 | 0.60653 | 0.7 | 1.6 |
| −1 | 0.36788 | 0.4 | 2.7 |
| −1.5 | 0.22313 | 0.2 | 4.5 |
| −2 | 0.13534 | 0.1 | 7.4 |
| . . . . . . | . . . . . . | . . . . . . | . . . . . . |
| −3 | 0.04979 | 0.05 | 20.1 |

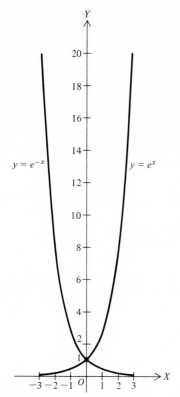

**Figure 6.2**

They are mirror images of one another, the mirror being the $Y$ axis. That is, they are symmetric with respect to the $Y$ axis.

In the next section we consider a number of applications of the exponential function in the fields of economics, biology, and physics.

## PROBLEMS 6.3

*In Probs. 1 to 10 obtain the values from Table II, Appendix B.*

| | | | | | | | | | |
|---|---|---|---|---|---|---|---|---|---|
| **1** | $e$ | **2** | $e^{-1}$ | **3** | $e^{-2.1}$ | **4** | $e^{3.4}$ | **5** | $e^{0.7}$ |
| **6** | $2e^{-2.6}$ | **7** | $1/e^{2.4}$ | **8** | $-1/e^{4.1}$ | **9** | $e^2 - e^{-2}$ | **10** | $e^{3.9}/e^{-0.8}$ |

*In Probs. 11 to 20 state those which define an exponential function.*

| | | | | | | | |
|---|---|---|---|---|---|---|---|
| **11** | $2x^2$ | **12** | $3^{3x}$ | **13** | $3(6.2)^x$ | **14** | $6(5)^x$ |
| **15** | $3^x + 4^x$ | **16** | $1^x$ | **17** | $0^x, 0 < x$ | **18** | $2/e^x$ |
| **19** | $4(4^x)$ | **20** | $e^x - e^{2x}$ | | | | |

In Probs. 21 to 34 make a table of values using $x = -2, -1, 0, 1,$ and $2,$ and sketch each in the interval $-2 \leq x \leq 2.$

| | | |
|---|---|---|
| **21** $y = 2e^{x/2}$ | **22** $y = 2e^{-x/2}$ | **23** $y = \frac{1}{2}e^{-x}$ |
| **24** $y = -e^{0.8x}$ | **25** $y = e^x + e^{-x}$ | **26** $y = e^x - e^{-x}$ |
| **27** $y = -2^x$ | **28** $y = 2^{-x}$ | **29** $y = (\frac{1}{2})^{-x}$ |
| **30** $y = (\frac{1}{2})^x$ | **31** $y = xe^x$ | **32** $y = xe^{-x}$ |
| **33** $y = -xe^x$ | **34** $y = -xe^{-x}$ | |

## 6.4    APPLICATIONS

In these days banks in the United States do a lot of talking about different ways of compounding interest on savings account deposits. Interest is expressed as a rate per 100 (per *cent*, from the Latin word *centum* meaning one *hundred*). Thus 5 percent, or 5%, of any quantity $S$ means $\frac{5}{100} \times S,$ or $(0.05)S.$ Suppose a bank pays 5 percent interest per year compounded yearly and suppose you deposit $100 with the bank. We shall call the original deposit the principal $P$; here $P = 100.$ One year later you will have in your account the original $100 plus $(0.05)(100),$ or a total of

$$\overset{P}{100} + \overset{\text{Interest}}{0.05(100)}$$

which we write in the form

$$100(1 + 0.05) = 105$$

Now if this new amount is left in the bank for another year, the deposit account will be a new principal of $105.

$$\overset{\text{New } P}{105} + \overset{\text{New interest}}{0.05(105)} = 110.25$$

Since interest is being paid on interest, we speak of compounding interest, or simply, *compound interest*. At the end of the third year the deposit account would amount to

$$110.25 + 0.05(110.25) = 115.76$$

rounding off to the nearest penny. And so on.

    We can readily generalize this and develop a formula which would give the amount at the end of $t$ years. So we deposit $P$ dollars and assume the rate of interest to be $r$ percent. At the end of 1 year we have our principal $P$ and $Pr$ dollars interest, a total of $A_1$ dollars (the subscript 1 indicating 1 year)

$$A_1 = P + Pr = P(1 + r)$$

The new principal is $P(1 + r)$, and the interest on it, at the end of the second year is $P(1 + r)r$, so the total 2-year amount $A_2$ is

$$A_2 = P(1 + r) + P(1 + r)r$$
$$= P(1 + r)(1 + r)$$
$$= P(1 + r)^2$$

For 3 years the total $A_3$ is

$$A_3 = P(1 + r)^2 + P(1 + r)^2(r)$$
$$= P(1 + r)^2(1 + r)$$
$$= P(1 + r)^3$$

For $t$ years,    $A_t = P(1 + r)^t$

Now interest rate is based on a full year, and the formula above is for annual compounding. The formula must be modified if the compounding is not on an annual basis.

For example, a principal of $100 compounded semiannually at an annual rate of 12 percent yields, at the end of 6 months (one period), the amount

$$\text{Principal} \quad \text{Interest}$$
$$A = 100 + 100(\tfrac{0.12}{2})$$
$$= 100(1 + \tfrac{0.12}{2}) = \$106$$

and at the end of two periods, or 1 year,

$$\text{New principal} \quad \text{Interest}$$
$$A = 100(1 + \tfrac{0.12}{2}) + 100(1 + \tfrac{0.12}{2})(\tfrac{0.12}{2})$$
$$= 100(1 + \tfrac{0.12}{2})^2 = 100(1.06)^2 = \$112.36$$

For 3 years the amount is

$$A = 100(1 + \tfrac{0.12}{2})^{2 \times 3} = 100(1.06)^6 = \$141.85$$

For compounding quarterly at 10 percent annual rate for one-quarter the amount is, for a principal of $P$ dollars,

$$A = P(1 + \tfrac{0.10}{4}) = P(1.025)$$

and for a year (4 periods),

$$A = P(1 + \tfrac{0.10}{4})^4 = P(1.1038)$$

and for 3 years,

$$A = P(1 + \tfrac{0.10}{4})^{4 \times 3}$$
$$= P(1 + \tfrac{0.10}{4})^{12} = P(1.3449)$$

The formula for $A$, with principal $P$ and an annual interest rate of

$r$ percent with $n$ compounding periods per year and for $t$ years is

$$A = P\left(1 + \frac{r}{n}\right)^{nt} \tag{1}$$

This formula holds for any real number $t$, even for $t$ negative, when we regard $A$ as the value of $P$ at a date $t$ years in the past. For fixed $P$, $r$, and $n$, this is of the form $A = ka^{bt}$, and so defines an exponential function with independent variable $t$.

Currently the most popular period is daily compounding. Most banks use 360 days for a year. At 5 percent compounded daily for 1 year, $100 grows into

$$A = \$100(1 + \tfrac{0.05}{360})^{360} = \$100(1 + 0.00013889)^{360}$$
$$= \$100(1.00013889)^{360} = \$105.126751$$

correct to five decimals, which rounds off to $105.13. This is not much of an increase over quarterly or even annual compounding. The computations were made with a hand calculator.

Some banks are offering even better than daily compounding: *continuous compounding*. This is the ultimate. Your money grows continuously, not by the day, not by the hour, minute, or second, but continuously, whatever that means. It does indeed mean something but to give an analysis of it would involve us in the theory of "limits," the mathematics of which is beyond the scope of this book. However, we can gain some understanding of the problem by examining what is happening to the amount $A$, for a fixed $r$ and $t$, as the number $n$ of compounding periods per year increases. It will be sufficient to consider the special case of formula (1) of this section when $r = 100$ percent and $t = 1$. Thus

$$A = P\left(1 + \frac{1}{n}\right)^{n} \tag{2}$$

and we compute $A$ for each of the following values of $n$.

**a** $n = 1$ (year, 360 days),

$$A = P(1 + \tfrac{1}{1})^{1} = P(2)$$

**b** $n = 12$ (compounding monthly),

$$A = P(1 + \tfrac{1}{12})^{12} = P(2.61303528)$$

**c** $n = 360$ (compounding daily),

$$A = P(1 + \tfrac{1}{360})^{360} = P(2.71451624)$$

**d** $n = 8640$ (compounding hourly),

$$A = P(1 + \tfrac{1}{8640})^{8640} = P(2.71813057)$$

**e**   $n = 518{,}400$ (compounding every minute),

$$A = P(1 + \tfrac{1}{518{,}400})^{518{,}400} = P(2.71826443)$$

**f**   $n = 31{,}104{,}000$ (compounding every second),

$$A = P(1 + \tfrac{1}{31{,}104{,}000})^{31{,}104{,}000} = P( \ ? \ )$$

and this is beyond the capacity of our hand calculator.

Note that while the quantity in parentheses, $1 + 1/n$, is getting smaller as $n$ gets larger, it is being raised to a larger power $n$ and in cases a to e, the net result is an increase in the value of $A$.  Are the values of $(1 + 1/n)^n$ as $n$ increases approximations to some number? What we really would like to know is what, if anything, would $(1 + 1/n)^n$ approach with ever-increasing $n$.  In technical language this translates into the question: What is the limiting value of $(1 + 1/n)^n$ as $n$ increases without bound?  This we have said we cannot explain at this level, but the answer is this limit is a number you have already met: $e = 2.71828$, correct to five decimal places.  Earlier we stated that $e^x$ was given by an infinite series [equation (1), Sec. 6.3].  When $x = 1$, this series sums to $e$, and so now we have two definitions of the very important number $e$.

With a little algebraic work with exponents we can rewrite formula (1) of this section in another form.  In this formula we set $r/n = 1/m$ from which it follows that $n = mr$.  We get

$$A = P\left(1 + \frac{r}{n}\right)^{nt} = P\left(1 + \frac{1}{m}\right)^{mrt}$$
$$= P\left[\left(1 + \frac{1}{m}\right)^{m}\right]^{rt}$$

Now with $r$ constant, as $n$ increases so does $m$, and therefore as $m$ increases without bound, the limiting value of $(1 + 1/m)^m$ is $e$, and so for continuous compounding, we have

$$A = Pe^{rt} \tag{2}$$

and for a given principal $P$ and an interest rate $r$, it is a simple matter to compute $A$ for $t$ years using Table II, Appendix B.  Here are some examples.

$$P = 100 \qquad r = 0.04 \qquad t = 2 \qquad rt = 0.08$$
and $\quad A = 100e^{rt} = 100e^{0.08} = 100(1.0833) = 108.33$

$$P = 1000 \qquad r = 0.10 \qquad t = 4 \qquad rt = 0.40$$
and $\quad A = 1000e^{rt} = 1000e^{0.40} = 1000(1.4918) = 1491.8$

We would need a larger table of values of $e$ to get the number of cents.

$$P = 100 \qquad r = 0.09 \qquad t = 3 \qquad rt = 0.27$$

and $\qquad A = 100e^{rt} = 100e^{0.27}$

and we must interpolate because of the small size of our $e$ table. Large tables exist. In our table we find $e^{0.20} = 1.2214$ and $e^{0.30} = 1.3499$. Our number, $e^{0.27}$ lies in between. We spread the computation out so we can see it the better:

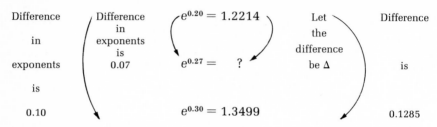

Now the ratio of the differences on the left $0.07/0.10$ must be equal to the ratio of the differences on the right (in the same order), namely $\Delta/0.1285$. Thus:

$$\frac{0.07}{0.10} = \frac{\Delta}{0.1285}$$

$$\Delta = \frac{(0.07)(0.1285)}{0.10}$$

$$= 0.0900$$

This is to be added to 1.2214 to yield the answer:

$$e^{0.27} = 1.3114$$

and $\qquad\qquad\qquad\qquad A = 131.14$

▶ **Illustration 6.4**

Problems similar to the above arise in biology, where each of $P$ cells in a given culture splits into two cells in a certain time $t$, and this leads to an exponential function. To see this let the number of cells initially (at time $T = 0$) be, say, 1. The number of cells $y$ increases as follows:

$$\begin{aligned}
&\text{At } T = 0 \qquad y = 1 = 2^0 \\
&\text{At } T = 1 \qquad y = 2 = 2^1 \\
&\text{At } T = 2 \qquad y = 4 = 2^2 \\
&\text{At } T = 3 \qquad y = 8 = 2^3 \\
&\qquad \cdots\cdots\cdots\cdots\cdots \\
&\text{At } T = t \qquad y = 2^t
\end{aligned}$$

and this defines an exponential function. It can be expressed in terms of base $e$ instead of base 2 as you will learn in Chap. 8. ◀

◆ **Illustration 6.5**

Suppose that the number of bacteria in a culture at time $t$ is given by

$$y = N_0 e^{5t}$$

**a** What is the number present at time $t = 0$? The answer is found by setting $t = 0$. Thus the number at zero time is $y = N_0 e^{5 \times 0} = N_0 e^0 = N_0$.
**b** When is the colony double its initial size? We must find $t$ when $y = 2N_0$.

$$2N_0 = N_0 e^{5t}$$

or

$$2 = e^{5t}$$

and from Table II, Appendix B,

$$5t = 0.70 \quad \text{approximately}$$
$$t = 0.14 \quad \text{unit of time} \quad ◀$$

◆ **Illustration 6.6**

A colony of bacteria was doubling every minute from early morning until 12:00 noon when the number was N. At what time was the number N/2? (At 11:59 A.M., of course.) ◀

Here is an example from atomic physics.

◆ **Illustration 6.7**

It is known that if $N_0$ is the number of $\pi^0$ mesons ("pi zero mesons") generated at time $t = 0$, then $y = N_0 e^{-at}$ is the number at any subsequent time. If only $N_0/2$ are present when $t = 3 \times 10^{-16}$ second, find $a$. This is known as the half-life of $\pi^0$ mesons. ($10^{-16} = 1/10^{16} = 1/10,000,000,000,000,000$.) We need to solve for $a$ in

$$\tfrac{1}{2}N_0 = N_0 e^{-a \times 3 \times 10^{-16}}$$

We cannot use Table II, Appendix B, since it is not adequate. You will find the solution set in Prob. 9 of Sec. 8.4. ◀

**PROBLEMS 6.4**

In Probs. 1 and 2, $P = \$1$ and $r = 4$ percent. Write the equation for the amount A based on

**1**    Quarterly compounding for 6 months.  Compute $A$.

**2**    Daily compounding for 2 years (360 days per year).

*In Probs. 3 and 4 write the equation for the principal P that would yield $A = $100$ at 4 percent at the end of*

**3**    3 months compounding quarterly.  Compute $P$.

**4**    2 years (360 days per year) compounding daily.

**5**    Approximately how long would it take, with continuous compounding at 6 percent, for a principal $P$ to double?

**6**    Approximately how long would it take, with continuous compounding at 7 percent, for a principal $P$ to triple?

**7**    One dollar is placed in a savings account to accumulate at 5 percent interest compounded daily.  Write the formula for the accumulated amount at the end of 100 years (1 year $=$ 360 days).

**8**    On the day of the birth of his daughter a man deposits a certain amount of money in a savings account at 5.5 percent interest compounded daily.  He wishes this to accumulate to $1 million on her twenty-first birthday.  Write the formula for the amount deposited.

**9**    A person wishes to deposit $P$ dollars in an 8 percent savings account so that, with interest compounded quarterly, there will be $500 at the end of 6 months.  Determine $P$.

**10**    Find the principal $P$ if the amount $A$, compounded continuously for 2 years at 5 percent is $1000.

**11**    A loan shark is to be paid $135 at the end of 1 year on a loan of $100. Approximate what is the interest rate compounded continuously?

**12**    A loan shark is to be paid $300 at the end of 2 years on a loan of $200. Approximately what is the interest rate compounded continuously?

**13**    Radium decomposes according to the formula $y = k_0 e^{-0.038t}$, where $k_0$ is the initial amount, and where $y$ is the amount undecomposed at time $t$ (in centuries).  Write an equation that must be solved for $t$ in order to find the time when one-half of the original amount will remain.  This is known as the half-life of radium.

**14**    In a given experiment the electron density doubles from 10 A.M. every minute. At 12 noon it is $D$.  At what time was it $D/4$?

# 7

# INVERSE
# FUNCTIONS

## 7.1 ONE-TO-ONE FUNCTIONS

In Chap. 3 you learned that a function is a rule which assigns each
number in one set (called the domain) to one and only one number in
another set (called the range). The rule is usually described by indicating which number $f(x)$ in the range is related to each number $x$ in the
domain. Arrows can sometimes be used to communicate the relationship between the domain elements and range elements of a function:

$$x \rightarrow f(x)$$

as in Fig. 7.1.

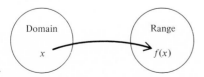

**Figure 7.1**

▶ **Illustration 7.1**

The equation $f(x) = 2x$ defines the function which assigns each real number $x$ in its domain to the real number $2x$ in its range. For example,

$$f(-1) = -2 \qquad f(\tfrac{3}{2}) = 3 \qquad f(61) = 122 \quad \blacktriangleleft$$

▶ **Illustration 7.2**

The expression $x \to x^2$ describes a function with domain equal to the real numbers and with range equal to the nonnegative real numbers. In particular,

$$
\begin{array}{l}
-1 \searrow \\
\phantom{-1}\quad 1 \\
1 \nearrow \\[4pt]
-2 \searrow \\
\phantom{-2}\quad 4 \\
2 \nearrow \\[4pt]
-\sqrt{2} \searrow \\
\phantom{-\sqrt{2}}\quad 2 \\
\sqrt{2} \nearrow \\[4pt]
0 \to 0 \quad \blacktriangleleft
\end{array}
$$

▶ **Illustration 7.3**

The following diagram defines a function.

$$
\begin{array}{l}
1 \to -1 \\
2 \to \phantom{-}5 \\
3 \to \phantom{-}3 \\
4 \to \phantom{-}6
\end{array}
$$

The domain of the function is $\{1, 2, 3, 4\}$ and the range of the function is $\{-1, 5, 3, 6\}$. ◀

▶ **Illustration 7.4**

Another function like the one in Illustration 7.3 is described by

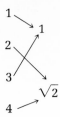

You determine the domain and range. ◀

We want to introduce you to one new property which some functions possess. You know that each x in the domain of a function is related to one and only one $f(x)$ in the range. But, as you can see in Illustrations 7.2 and 7.4, some functions relate more than one x to the same number in the range. For example, in Illustration 7.2, $f(-2) = f(2)$, that is, both $-2$ and 2 are assigned to 4; and in Illustration 7.4, $f(1) = f(3)$; that is, both 1 and 3 are assigned to 1. Can you see, on the other hand, that the functions in Illustrations 7.1 and 7.3 relate exactly one x to each number in the range? Functions like these latter ones are called *one-to-one*.

## DEFINITION 7.1

A one-to-one function is a function which assigns exactly one number in its domain to each number in its range. ◀

## ▶ Illustration 7.5

The function defined by $f(x) = 2$ for each real number x is not one-to-one, since it assigns many different x's to the same $f(x)$. ◀

## ▶ Illustration 7.6

The function defined by $f(x) = \frac{1}{2}x + 7$ is one-to-one. To prove this we need to show that the same $f(x)$ cannot result from two different x's. Suppose $f(x_1) = f(x_2)$. Then $\frac{1}{2}x_1 + 7 = \frac{1}{2}x_2 + 7$. From which we get $x_1 = x_2$. So we see that if $f(x_1) = f(x_2)$, then $x_1 = x_2$; i.e., the same $f(x)$ cannot come from different x's. So the function defined by $f(x) = \frac{1}{2}x + 7$ is one-to-one. ◀

## PROBLEMS 7.1

**1** Indicate which of the functions defined below is one-to-one. Show that the non-one-to-one functions are not one-to-one.

**a**

| $x$ | 1 | 2 | 3 | 4 | 5 |
|------|---|---|---|---|---|
| $f(x)$ | 5 | 4 | 3 | 2 | 1 |

**b**

| $x$ | $-2$ | $-1$ | 0 | 1 | 2 |
|------|------|------|---|---|---|
| $f(x)$ | 2 | 1 | 0 | 1 | 2 |

**c**

| $x$ | 1 | 7 | 8 | 16 | 100 |
|------|---|---|---|----|-----|
| $f(x)$ | $-1$ | 0 | 1 | 2 | 3 |

**d**

| $x$ | $\pi$ | $e$ | 10 | 7 | 0 |
|------|-------|-----|----|---|---|
| $f(x)$ | $-1$ | 0 | 1 | $-1$ | 2 |

**2** Indicate which of the functions defined below is one-to-one. Show that the non-one-to-one functions are not one-to-one.

**a**

| $x$ | $-7$ | 2 | 3 | 5 | 15 |
|------|------|---|---|---|----|
| $f(x)$ | 1 | 7 | 16 | $-1$ | 1 |

**b**

| $x$ | $\frac{1}{2}$ | $\frac{1}{3}$ | $\frac{1}{4}$ | $\frac{1}{5}$ | $\frac{1}{6}$ | $\frac{1}{7}$ |
|------|---------------|---------------|---------------|---------------|---------------|---------------|
| $f(x)$ | 2 | 3 | 4 | 5 | 6 | 7 |

**c**

| $x$ | $-1$ | 1 | $-2$ | 2 | $-3$ | 3 |
|------|------|---|------|---|------|---|
| $f(x)$ | 1 | $-1$ | 2 | $-2$ | 3 | $-3$ |

**d**

| $x$ | 1 | 2 | 3 | 4 | 5 | 6 |
|------|---|---|---|---|---|---|
| $f(x)$ | 5 | 4 | 3 | 2 | 1 | 5 |

**3** Indicate which of the functions defined below is one-to-one. Show that the non-one-to-one functions are not one-to-one.

**a** $f(x) = -\frac{1}{3}x$ for all $x$     **b** $f(x) = x^4$ for all $x$
**c** $f(x) = x^2$ for $x \geq 0$     **d** $f(x) = 1$ for all $x$

**4** Indicate which of the functions defined below is one-to-one. Show that the non-one-to-one functions are not one-to-one.

**a** $f(x) = 3x + 1$ for all $x$     **b** $f(x) = \sqrt{x}$ for $x \geq 0$
**c** $f(x) = x^6$ for all $x$     **d** $f(x) = x^2$ for $x \leq 0$

**5** Show that if $a \neq 0$, the function defined by $f(x) = ax + b$ is one-to-one.
**6** Show that a constant function whose domain has more than one element in it is not one-to-one.
**7** Make up two examples of one-to-one functions and two examples of non-one-to-one functions.

## 7.2    INVERSES

In looking for an inverse of a function, we are looking for a function which does the reverse of the function. If the function assigns 1 to 3, its inverse will assign 3 to 1. Can you see why we were concerned about one-to-one functions? Suppose that $f(1) = 3$ and $f(2) = 3$. What would the inverse of $f$ assign 3 to, 1 or 2? However, if $f$ is one-to-one, each $f(x)$ will have come from only one $x$, so that the inverse of $f$ will assign $f(x)$ to $x$. First we consider some examples before we present the formal definitions.

### ♦ Illustration 7.7

In Illustration 7.3 a one-to-one function was defined by the diagram below.

$$1 \to -1$$
$$2 \to \;\; 5$$
$$3 \to \;\; 3$$
$$4 \to \;\; 6$$

The inverse of this function is defined by a similar diagram.

$$-1 \to 1$$
$$5 \to 2$$
$$3 \to 3$$
$$6 \to 4 \quad ♦$$

### ♦ Illustration 7.8

In Illustration 7.1 a function $f$ was defined by the equation $f(x) = 2x$. Its inverse $f^{-1}$ is defined by the equation $f^{-1}(x) = x/2$. Let us see why. Note that $f$ assigns $x$ to $2x$. We want $f^{-1}$ to assign $2x$ back to $x$. Since $\frac{1}{2}(2x) = x$, we conclude that $f^{-1}$ is that function which assigns to each number one-half of that number. ♦

### ♦ Illustration 7.9

The functions in Illustration 7.2 defined by $f(x) = x^2$ do not have an inverse. Since $f(-2) = f(2) = 4$, there is no way to define $f^{-1}$. If $f^{-1}$ were defined, it would have to assign 4 to both 2 and $-2$, and we know that functions are not allowed to assign one number to two numbers. ♦

### DEFINITION 7.2

Let $f$ be a one-to-one function. The inverse of $f$ (designated by $f^{-1}$) is the function which assigns $f(x)$ to $x$ for each $f(x)$ in the range of $f$. The domain of $f^{-1}$ is the same as the range of $f$, and the range of $f^{-1}$ is the same as the domain of $f$. ◀

Referring to the various notations used for functions, we have

$$f^{-1}(f(x)) = x$$

If $y = f(x)$, $x = f^{-1}(y)$.

In each case you can see that the function $f^{-1}$ is reversing or undoing the function $f$. See Fig. 7.2.

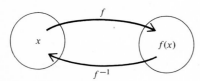

**Figure 7.2**

## PROBLEMS 7.2

**1** Define the inverse of each one-to-one function in Prob. 1 of Sec. 7.1.
**2** Define the inverse of each one-to-one function in Prob. 2 of Sec. 7.1.
**3** Define the inverse of each one-to-one function in Prob. 3 of Sec. 7.1.
**4** Define the inverse of each one-to-one function in Prob. 4 of Sec. 7.1.

## 7.3    FINDING AND GRAPHING INVERSES

Functions can be defined and referred to in many different ways. Frequently we have used an equation such as $f(x) = 3x - 5$ to define a function. Sometimes we have used $y = 3x - 5$. Letting $y = f(x)$ in this way is particularly convenient when graphing functions.

If a function assigns x to y:

$$\begin{array}{c} f \\ x \rightarrow y \end{array}$$

we know that its inverse assigns y back to x:

$$\begin{array}{c} f^{-1} \\ x \leftarrow y \end{array}$$

So if $f(x) = 3x - 5$ or $y = 3x - 5$ defines $f$, we can solve for x in terms of y by adding 5 to both sides of the equation and then dividing both sides by 3 to get

$$x = \frac{y + 5}{3}$$

From this we can conclude that

$$f^{-1}(y) = \frac{y + 5}{3}$$

But we are used to using x to designate the domain element of a function, so we can replace $y$ by $x$ to get

$$f^{-1}(x) = \frac{x + 5}{3}$$

(Do not let the changing of variables confuse you. The choice of letters is mostly habit, or convenience. We could have used $r$ and $s$ in place of $x$ and $y$, or any other letters for that matter.) Another way to get the same result is to interchange $x$ and $y$ initially as follows. We have $y = 3x - 5$. Interchange $x$ and $y$ to get $x = 3y - 5$. Solve for $y$ in terms of $x$ to get $y = (x + 5)/3$ which is an equation defining the inverse.

▶ **Illustration 7.10**

Suppose $f(x) = \frac{1}{8}x - \frac{3}{4}$ defines the function $f$. To find $f^{-1}$, we just write $y = \frac{1}{8}x - \frac{3}{4}$. Then we interchange $x$ and $y$ to get $x = \frac{1}{8}y - \frac{3}{4}$. Then we solve for $y$ in terms of $x$:

$$y = \frac{x + \frac{3}{4}}{\frac{1}{8}} \qquad \text{or} \qquad y = 8x + 6$$

This equation defines $f^{-1}$ which can also be rewritten as

$$f^{-1}(x) = 8x + 6 \quad ◀$$

▶ **Illustration 7.11**

This technique will even help you spot non-one-to-one functions.
    Suppose that $f(x) = x^2 + 3x + 2$. We write this as $y = x^2 + 3x + 2$. To try to find $f^{-1}$ we interchange $x$ and $y$:

$$x = y^2 + 3y + 2$$

Notice that this is a quadratic equation in the variable $y$. So to solve for $y$ in terms of $x$ we write this as

$$y^2 + 3y + (2 - x) = 0$$

and apply the quadratic formula, treating $2 - x$ as the constant term $c$.

$$y = \frac{-3 \pm \sqrt{3^2 - 4(1)(2 - x)}}{2}$$

$$= \frac{-3 \pm \sqrt{1 + 4x}}{2}$$

You can see from this that for $1 + 4x > 0$, there are two $y$'s for each $x$, namely,

$$\frac{-3 + \sqrt{1 + 4x}}{2} \quad \text{and} \quad \frac{-3 - \sqrt{1 + 4x}}{2}$$

This situation reflects the fact that the original function was not one-to-one, so it does not have an inverse function.  ◀

The fact that the equation for $f^{-1}$ can be found by interchanging $y$ and $x$ in the equation for $f$ has implications for graphing. In fact, you will see that if you know the graph of $f$, you will be able to sketch the graph of $f^{-1}$ without referring to any equation.

Before considering the graphs of inverses, we want to look at the effect of interchanging $x$ and $y$ in an ordered pair in Fig. 7.3. You can

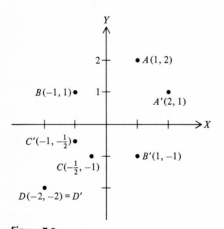

Figure 7.3

see that for each of the points $A(1, 2)$, $B(-1, 1)$, $C(-\frac{1}{2}, -1)$, $D(-2, -2)$ we have switched the order of the coordinates to get $A'(2, 1)$, $B'(1, -1)$, $C'(-1, -\frac{1}{2})$, and $D'(-2, -2)$. One way of describing the configuration is to say that switching the order of the coordinates reflects the point on the line $y = x$ (Fig. 7.4). For any point $P(a, b)$, the line $y = x$ is the perpendicular bisector of the line from $P(a, b)$ to $P'(b, a)$ (Fig. 7.5).

Let us verify that this is always true.

---

**THEOREM 7.1**    *The line $y = x$ is the perpendicular bisector of the line segment joining $P(a, b)$ to $P'(b, a)$, assuming $a \neq b$.*

---

**PROOF**

To show that the line is perpendicular to $y = x$, it will suffice to show that the line has slope equal to $-1$. This is true since the line $y = -x$ whose slope is $-1$ is perpendicular to $y = x$, so that any other line with slope $-1$ will be parallel to $y = -x$, and therefore perpendicular to $y = x$. To show that $y = x$ bisects the segment we will show that $P$ and $P'$ are the same distance from $y = x$ (Fig. 7.5).

The slope of the line from $P$ to $P'$ is

$$\frac{b-a}{a-b} = -\frac{b-a}{b-a} = -1$$

So the lines are perpendicular. Where do they intersect? The equation of the line $\overline{PP'}$ is $y - b = -1(x - a)$. Solving this equation simultaneously with $y = x$ we substitute $y = x$ into $y - b = -1(x - a)$ to get

$$y - b = -y + a$$

or

$$2y = a + b$$

So

$$y = \frac{a+b}{2}$$

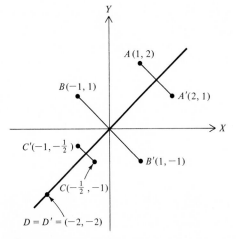

Figure 7.4

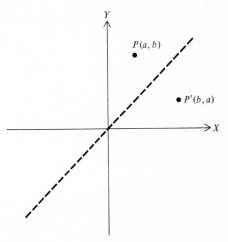

Figure 7.5

But we know that $y = x$ at the point of intersection, so we conclude that $PP'$ and $y = x$ intersect at

$$\left(\frac{a+b}{2}, \frac{a+b}{2}\right)$$

Finally, the distance of $P$ from the intersection is

$$\sqrt{\left(a - \frac{a+b}{2}\right)^2 + \left(b - \frac{a+b}{2}\right)^2}$$

while the distance of $P'$ from the intersection is

$$\sqrt{\left(b - \frac{a+b}{2}\right)^2 + \left(a - \frac{a+b}{2}\right)^2}$$

Since these are equal, we have that $y = x$ bisects $PP'$.

Earlier in the section we saw that $y = f(x)$ if and only if $x = f^{-1}(y)$. So that whenever $(a, b)$ lies on the graph of $f$, $(b, a)$ lies on the graph of $f^{-1}$, and vice versa. So that to find the graph of $f^{-1}$, one need only find all the $(b, a)$'s so that $(a, b)$ is on the graph of $f$. But Theorem 7.1 tells us that to find $(b, a)$ one need only find the mirror image of $(a, b)$ in the line $y = x$. So we have the following rule.

**Rule 7.1**
To find the graph of $f^{-1}$, you need only find the mirror image of the graph of $f$ in the line $y = x$.

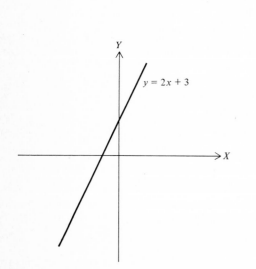

$y = 2x + 3$

**Figure 7.6**

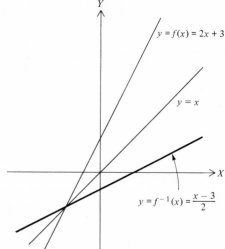

$y = f(x) = 2x + 3$

$y = x$

$y = f^{-1}(x) = \dfrac{x-3}{2}$

**Figure 7.7**

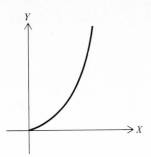

**Figure 7.8**

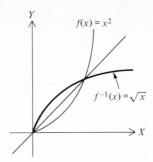

**Figure 7.9**

▶ **Illustration 7.12**

Graph $f^{-1}(x)$, where $f(x) = 2x + 3$. First the graph of $f$ is shown in Fig 7.6. So to find the graph of $f^{-1}$ you need only "reflect" the graph above in the line $y = x$ (Fig. 7.7). ◀

▶ **Illustration 7.13**

Graph $f^{-1}(x)$, when $f(x) = x^2$ for $x \geq 0$. The graph of $y = x^2$ for $x \geq 0$ is shown in Fig. 7.8. "Reflecting" in the line $y = x$ we get Fig. 7.9. ◀

## PROBLEMS 7.3

**1** Use the technique of interchanging $x$ and $y$ to find $f^{-1}$ for each of the following $f$'s.

    **a** $f(x) = 2x^3 + 4$      **b** $f(x) = 4 - 3x$

    **c** $f(x) = \dfrac{1}{x+3}$      **d** $f(x) = x^2 + 3$

**2** Use the technique of interchanging $x$ and $y$ to find $f^{-1}$ for each of the following $f$'s.

    **a** $f(x) = \dfrac{3x}{2} - 5$      **b** $f(x) = x^5 - 2$

    **c** $f(x) = \dfrac{x}{x+2}$      **d** $f(x) = (x-1)(x+2)$

**3** Graph $f^{-1}$ for each of the following functions by graphing $f$ and then using Rule 7.1.

    **a** $f(x) = -2x + 4$      **b** $f(x) = x^3$
    **c** $f(x) = x^2, x \leq 0$      **d** $f(x) = 2^x$

**4** Graph $f^{-1}$ for each of the following functions by graphing $f$ and then using Rule 7.1.

**a** $f(x) = \dfrac{3-x}{7}$       **b** $f(x) = x^5$

**c** $f(x) = x^4,\ x \leq 0$       **d** $f(x) = (\tfrac{1}{2})^x$

**5** Show that if the graphs of $f$ and $f^{-1}$ intersect, any intersection must be on the line $y = x$.

**6** Show that if the graph of $f$ intersects the line $y = x$, then the graphs of $f$ and $f^{-1}$ will intersect at that point of intersection.

## 7.4   SOME ALGEBRA WITH FUNCTIONS

In the preceding three sections we have been learning about one-to-one functions and inverses, and we have been doing some specific computation with specific functions. In this section we will observe a few general relationships involving functions.

First you will recall that in Chap. 3 we introduced the composition of functions with the following definition.

### DEFINITION 7.3

If $f$ and $g$ are functions so that the range of $f$ is contained in the domain of $g$, then the function $g \circ f$ is defined by $g \circ f(x) = g(f(x))$ for each $x$ in the domain of $f$; and $g \circ f$ is read, "g composed with $f$," or "composition of $g$ and $f$." ◀

### DEFINITION 7.4

The identity function is defined to be the function $E$ defined by $E(x) = x$ for all $x$. ◀

From Definition 7.2 we easily get

### THEOREM
### 7.2          Let $f$ be a one-to-one function.   Then $f^{-1} \circ f = E$.

**PROOF**

By definition of $f^{-1}$, $f^{-1}(f(x)) = x$. But $f^{-1} \circ f(x) = f^{-1}(f(x)) = x = E(x)$. So $f^{-1} \circ f = E$.

### THEOREM
### 7.3          If $f$ is a one-to-one function, $(f^{-1})^{-1} = f$.

**PROOF**
Speaking loosely, $(f^{-1})^{-1}(x)$ is that number that $f^{-1}$ assigns to $x$. But $f^{-1}(f(x)) = x$ by Theorem 7.2. So $(f^{-1})^{-1}(x) = f(x)$; that is, $(f^{-1})^{-1} = f$.

**THEOREM 7.4**    For any function $f$, $E \circ f = f \circ E = f$.

Do you see any analogy between function composition and the product of numbers? The function $E$ plays a role like the role of 1. $f^{-1} \circ f = E$; $E \circ f = f \circ E = f$; $(f^{-1})^{-1} = f$. There follow two more analogous results. Other similarities and differences between composition of functions and multiplication of numbers will be developed in the exercises.

**THEOREM 7.5**    Composition of functions is associative, that is,

$$f \circ (g \circ h) = (f \circ g) \circ h$$

**PROOF**
$$f \circ (g \circ (h))(x) = f(g \circ h(x)) = f(g(h(x)))$$
$$(f \circ g) \circ h(x) = f \circ g(h(x)) = f(g(h(x)))$$
So    $$f \circ (g \circ h) = (f \circ g) \circ h$$

**THEOREM 7.6**    If $f$ and $g$ are one-to-one functions whose domains equal the set of real numbers, then there exists a one-to-one function $h$ so that $h \circ f = g$.

**PROOF**
Let $h = g \circ f^{-1}$. It is left as an exercise for you to show that it works.

## PROBLEMS 7.4

**1**  Let $f(x) = 3x + 7$ and $g(x) = (4 - 2x)/3$. Find $h$ so that $h \circ f = g$.
**2**  Let $f(x) = x + 1$ and $g(x) = 3 - x$. Find $h$ so that $h \circ f = g$.
**3**  Prove that the composition of one-to-one functions is not commutative. That is, find one-to-one functions $f$ and $g$ so that $f \circ g \neq g \circ f$.
**4**  Prove that if $f$, $g$, and $h$ are one-to-one functions and $f \circ g = h \circ g$, then $f = h$.

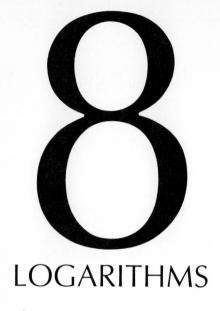

# LOGARITHMS

## 8.1 THE LOG FUNCTION

Exponential functions were introduced in Chap. 6. The concept of the inverse of a function was introduced in Chap. 7. In this chapter we will introduce inverses of exponential functions, which are called *logarithms* or *log functions*. Log functions are useful for solving problems involving exponential functions. Certain real-world phenomena can be described using logarithms, and logarithms arise in solutions to problems in the physical sciences. Traditionally, logarithms have been a useful tool in certain complex computations. This latter role will probably be diminished by the widespread use of hand calculators.

You will recall that for $b > 0$ and $b \neq 1$ the function $f(x) = b^x$ is a one-to-one function with domain equal to the whole set of real numbers and range equal to the positive real numbers. Since these exponential functions are one-to-one, they have inverses. These inverses are called logarithms. Since the range of the exponential functions is the positive reals, the domain of the logarithm functions is the positive reals. Moreover, since the domain of the exponential functions is the set of all real numbers, the range of the logarithm functions is the set of all real numbers. We now formally define the logarithm function.

### DEFINITION 8.1

The *logarithm function* with base $b$ ($b \neq 1$, $b > 0$) is the inverse of the exponential function with base $b$. It is the function which relates each $x > 0$ to a $y$ ($= \log_b x$) according to the following rule:

$$y = \log_b x \quad \text{if and only if} \quad x = b^y. \quad \blacktriangleleft$$

Before presenting properties of the logarithm function, we will present some illustrations of this definition.

▶ **Illustration 8.1**

$$\log_2 8 = 3$$

This is read "the log to the base 2 of 8 equals 3." Why? Because $2^3 = 8$.

$$\log_2 \tfrac{1}{8} = -3 \quad \text{since } 2^{-3} = \tfrac{1}{8}$$

Try $\log_2 x$ for $x = 2$, 32, 1, and $\tfrac{1}{2}$. Did you get 1, 5, 0, and $-1$? ◀

▶ **Illustration 8.2**

$$\log_{10} 100 = 2 \quad \text{since } 10^2 = 100$$
$$\log_{10} 0.001 = -3$$
$$\log_{10} 1 = 0$$
$$\log_{10} 10^n = n \quad \blacktriangleleft$$

▶ **Illustration 8.3**

$$\log_{1/3} 9 = -2 \quad \text{since } (\tfrac{1}{3})^{-2} = 9$$
$$\log_{1/3} \tfrac{1}{27} = 3$$

You make up some problems of the form $\log_{1/3} x = y$. ◀

You can rightfully accuse us of having chosen only very special $b$'s and $x$'s in the above illustrations of $\log_b x$. For example, what about $\log_2 10$ and $\log_3 5.3$, and what is $\log_{10} \sqrt{2}$? Each of these logs is also defined. In fact, each is an irrational number that you will learn how to approximate. The approximation of logarithms will be discussed in Sec. 8.3.

Some further insight into logarithms can be gained from studying their graphs. You know what the graphs of exponential functions look like (Fig. 8.1). You also know that the graph of the inverse of a function is the reflection or mirror image of the graph of the function in the line y = x. See Fig. 8.2. We can use this knowledge about the graphs of exponentials and of inverses to graph logarithms (Fig. 8.3).

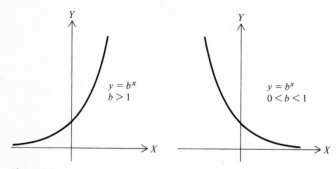

Figure 8.1

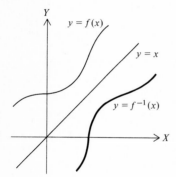

Figure 8.2

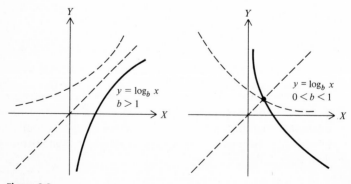

Figure 8.3

The exact shape of the graph of $\log_b x$ depends, of course, on the number $b$. The graphs drawn here are for $b = 2$ and $b = \frac{1}{2}$.

Before getting to the problems for this section we will summarize some facts about logarithms.

### Facts about Logarithms

The function $f(x) = \log_b x$ is the inverse of the function $g(x) = b^x$, so that $y = \log_b x$ if and only if $b^y = x$.

The function $\log_b x$ is defined for any positive real number $b$ except 1.

The function $\log_b x$ is defined for all $x > 0$. That is, the domain of $\log_b$ is the set of positive real numbers.

From looking at the graph of $\log_b x$ we see that if $b > 1$, $\log_b x$ is increasing; that is, if $x_1 > x_2$, $\log_b x_1 > \log_b x_2$. If $0 < b < 1$, $\log_b x$ is decreasing.

The range of $\log_b$ is the set of all real numbers.

It is also worth noting that logarithms are one-to-one functions. That is, if $\log_b x = \log_b y$, then $x = y$. This follows either from the fact that logs are inverses of one-to-one functions or from the fact that logs are increasing or decreasing.

## PROBLEMS 8.1

**1** Let $f(x) = \log_4 x$. Find each of the following.

    **a** $f(\frac{1}{16})$    **b** $f(1)$    **c** $f(2)$    **d** $f(\sqrt{2})$

**2** Let $g(x) = \log_9 x$. Find each of the following.

    **a** $g(\frac{1}{81})$    **b** $g(1)$    **c** $g(3)$    **d** $g(\sqrt{3})$

**3** Further sharpen your skill with the logarithm function by finding each of the following.

    **a** $\log_{1/3} 9$      **b** $\log_{2/3} \frac{3}{2}$      **c** $\log_{\sqrt{2}} 2$
    **d** $\log_{10} 0.0001$    **e** $\log_{10} 10{,}000$

**4** Find each of the following.

    **a** $\log_{1/4} 16$     **b** $\log_{10} 0.1$    **c** $\log_{0.1} 10$
    **d** $\log_{17} \sqrt{17}$    **e** $\log_{32} \frac{1}{2}$

**5** Explain why each of the following is not defined.

    **a** $\log_1 2$    **b** $\log_{10} 0$

**6** Find

    **a** $\log_b b^x$    **b** $(b)^{\log_b x}$

Explain your answers in terms of inverse functions.

**7** Fill in the following tables and graph the corresponding function.

**a**

| x | 0.01 | 0.1 | 1 | $\sqrt{10}$ | 10 | $(10)^{3/2}$ | 100 |
|---|---|---|---|---|---|---|---|
| $\log_{10} x$ | | | | | | | |

**b**

| x | $\frac{1}{27}$ | $\frac{1}{9}$ | $\frac{1}{3}$ | 1 | $\sqrt{3}$ | 3 | 9 | 27 |
|---|---|---|---|---|---|---|---|---|
| $\log_{1/3} x$ | | | | | | | | |

## 8.2 PROPERTIES OF LOGARITHMS

What is so special about logarithms? Presumably, each of the functions that we focus on has some features or properties which make it worthy of attention. We already know that logarithm functions are inverses of exponential functions. We know what the domain and range of each logarithm is. Here we prove three general theorems which describe extremely useful properties of logarithms. The proof of each theorem relies heavily on the fact that logarithms are inverses of exponentials which have certain properties. You should assume throughout that $b \neq 1$, $b > 0$, and that $x > 0$ and $y > 0$.

---

**THEOREM 8.1**

$$\log_b xy = \log_b x + \log_b y$$

**PROOF**

In order to take advantage of what we know about exponentials, we let $p = \log_b x$ and $q = \log_b y$. Then from the definition of logarithm we can write $x = b^p$ and $y = b^q$. This theorem is about $xy$, so we write

$$xy = (b^p)(b^q)$$
$$= b^{p+q} \quad \text{according to Theorem 6.1}$$

But since $xy = b^{p+q}$, we get $\log_b xy = p + q$. But $p = \log_b x$ and $q = \log_b y$, so we have $\log_b xy = \log_b x + \log_b y$, which is what we wanted to prove.

---

**THEOREM 8.2**

$$\log_b x^r = r \log_b x$$

**PROOF**

Again we let $p = \log_b x$, from which we get $x = b^p$. This theorem is about $x^r$, so we write

$$x^r = (b^p)^r$$
$$= b^{pr} \qquad \text{according to Theorem 6.4}$$
$$= b^{rp}$$

Since $x^r = b^{rp}$, $\log_b x^r = rp$. But $p = \log_b x$, so $\log_b x^r = r \log_b x$, which is what we set out to prove.

---

**THEOREM 8.3**

$$\log_b \frac{x}{y} = \log_b x - \log_b y$$

**PROOF**

One could prove this theorem as we did the other two by referring to Theorem 6.1 and to the fact that $b^p/b^q = b^{p-q}$. Instead we will use the above theorems for an easy proof.

$$\log_b \frac{x}{y} = \log_b (x \cdot y^{-1})$$
$$= \log_b x + \log_b y^{-1} \qquad \text{by Theorem 8.1}$$
$$= \log_b x + (-1) \log_b y \qquad \text{by Theorem 8.2}$$
$$= \log_b x - \log_b y$$

---

▶ **Illustration 8.4**

$$\log_2 8 = \log_2 2^3 = 3 \log_2 2 = 3 \cdot 1 = 3$$
$$\log_2 4 = \log_2 (2 \cdot 2) = \log_2 2 + \log_2 2 = 1 + 1 = 2$$
$$\log_2 0.5 = \log_2 \tfrac{1}{2} = \log_2 1 - \log_2 2 = 0 - 1 = -1$$

You already could have done any of the above without the theorems. But it is nice to see that new information is consistent with old. ◀

---

▶ **Illustration 8.5**

Now we want to find $\log_{10} r$ for any $r > 0$. This is a very important example for future consideration. First we need to notice a couple of facts about positive real numbers:

$$129 = 1.29 \times 100 = 1.29 \times 10^2$$
$$0.04397 = 4.397 \times 0.01 = 4.39 \times 10^{-2}$$
$$94{,}361 = 9.4361 \times 10^4$$
$$0.000279 = 2.79 \times 10^{-4}$$

In fact if $r$ is any positive real number, $r = t \times 10^n$, where $1 \leq t < 10$ and $n$ is an integer. And if $r = t \times 10^n$,

$$\log_b r = \log_b (t \times 10^n) = \log_b t + \log_b 10^n$$
$$= \log_b t + n \log_b 10$$

Thus, if we take $b = 10$, we get

$$\log_b 10 = \log_{10} 10 = 1 \quad \text{and} \quad \log_{10} r = n + \log_{10} t$$

So to find the $\log_{10}$ of any positive real number, you need only know the $\log_{10}$ of positive real numbers between 1 and 10. This illustration is the basis for the use of logarithms in computation. ◄

● **Illustration 8.6**

Here are three illustrations of how logarithms can be used in algebraic manipulations.

    **a**  If $\log_5 x = -3$, then $x = 5^{-3} = \frac{1}{125}$.

    **b**  If $\log_{10} x - \log_{10} (x + 1) = \log_{10} 2$, then $\log_{10} x/(x + 1) = \log_{10} 2$.
It follows since log functions are one-to-one that

$$\frac{x}{x + 1} = 2$$
$$x = 2(x + 1)$$

and so
$$x = -2$$

    **c**  If $5^{2x+1} = 25$, $\log_5 5^{2x+1} = \log_5 25$. But $\log_5 5^{2x+1} = (2x + 1) \log_5 5 = 2x + 1$. And $\log_5 25 = 2$. So $2x + 1 = 2$, and $x = \frac{1}{2}$. ◄

## PROBLEMS 8.2

    **1**  Compute each of the following in two different ways, one directly using the inverse relationship between logs and exponentials and the other using Theorems 8.1, 8.2, and 8.3.

    **a**  $\log_2 \frac{1}{32}$    **b**  $\log_2 0.25$    **c**  $\log_8 64$

    **2**  As in Prob. 1 compute each of the following in two different ways.

    **a**  $\log_{0.25} 16$    **b**  $\log_5 0.2$    **c**  $\log_{\sqrt{2}} 0.5$

    **3**  Prove that $-\log x = \log 1/x$ in two different ways — one directly using the inverse relationship between logs and exponentials and the other using Theorems 8.2 and 8.3.

    **4**  Express each of the following numbers in the form $t \times 10^n$, where $1 \le t < 10$ and $n$ is an integer.

    **a**  539        **b**  421.32      **c**  0.000391

    **d**  789,146.2    **e**  $\frac{1}{4}$        **f**  7.329

**5** Express each of the following numbers in the form $t \times 10^n$, where $1 \le t < 10$ and $n$ is an integer.

**a** 27.32 **b** 0.00761 **c** 4,932,651.1

**d** $\frac{1}{5}$ **e** 4.713 **f** 17

**6** Express $\log_{10}$ of each of the following numbers in the form $n + \log_{10} t$, where $1 \le t < 10$ and $n$ is an integer.

**a** 1.631 **b** 793.2 **c** 0.00071

**7** Express $\log_{10}$ of each of the following numbers in the form $n + \log_{10} t$, where $1 \le t < 10$ and $n$ is an integer.

**a** $\frac{1}{10}$ **b** 893.6 **c** 0.00372

**8** Solve each of the following.
  **a** If $4 = \log_2 x$, what is $x$?
  **b** If $-2 = \log_3 x$, what is $x$?
  **c** If $\frac{1}{2} = \log_{1/9} x$, what is $x$?
**9** Solve each of the following.
  **a** If $-3 = \log_{10} x$, what is $x$?
  **b** If $2 = \log_{10} x$, what is $x$?
  **c** If $0 = \log_{10} x$, what is $x$?
**10** Solve each of the following equations for $x$.
  **a** $\log_{10} x - \log_{10} (x - 3) = \log_{10} 4$
  **b** $\log_2 x^2 - \log_2 (3x - 2) = 0$
  **c** $\frac{1}{2} \log_7 x = \log_7 4$
**11** Solve each of the following equations for $x$.
  **a** $x \log_{10} 3 = \log_{10} 9$
  **b** $\log_2 x + \log_2 (x - 2) = \log_2 8$
  **c** $7^{\log_7 x} = 1$

## 8.3 COMPUTING LOGARITHMS

So far we have computed $\log_b x$ only for very special $b$'s and $x$'s. Unfortunately, many of the applications of logarithms involve less convenient $b$'s and $x$'s. In this section you will learn how to use Table III, Appendix B to compute $\log_b x$ for any positive real number $b$ besides 1 and for any positive real number $x$. Rather than try to describe the process of computing $\log_b x$ completely, we will illustrate and then summarize the process.

### ◆ Illustration 8.7

Find $\log_{10} 956$.

In Sec. 8.2 you saw that we can write 956 as $9.56 \times 10^2$. So that

$$\log_{10} 956 = \log_{10} (9.56 \times 10^2)$$
$$= \log_{10} 9.56 + \log_{10} 10^2$$
$$= \log_{10} 9.56 + 2$$

Now, looking at Table III, Appendix B, we see that $\log_{10} 9.56 = 0.9805$. So we conclude that

$$\log_{10} 956 = 0.9805 + 2$$
$$= 2.9805 \quad \blacklozenge$$

### ◆ Illustration 8.8

Find $\log_{10} 0.0168$.

Again, we rewrite 0.0168 as $1.68 \times 10^{-2}$. So that,

$$\log_{10} 0.0168 = \log_{10} (1.68 \times 10^{-2})$$
$$= \log_{10} 1.68 + \log_{10} 10^{-2}$$
$$= \log_{10} 1.68 + (-2)$$

Returning to Table III, Appendix B, we see that $\log_{10} 1.68 = 0.2253$. Thus we conclude that

$$\log 0.0168 = 0.2253 - 2$$
$$= -1.7747 \quad \blacklozenge$$

Summarizing what we have done so far, we see that Table III, Appendix B contains $\log_{10} r$ for $1 \leq r < 10$, which can be expressed with two decimal places. So that when confronted with any three-digit $x$, we express it as $x = r \times 10^n$, where $1 \leq r < 10$ and where $n$ is an integer.

Then we write

$$\log_{10} x = \log_{10} (r \times 10^n)$$
$$= \log_{10} r + \log_{10} 10^n$$
$$= \log_{10} r + n$$

Then we look up $\log_{10} r$ in Table III, Appendix B, and finally compute $\log_{10} x$.

We have so far restricted ourselves to three-digit $x$'s and to base 10. We will now learn to deal with $x$'s with more than three digits. Then we will deal with bases other than 10.

### ◆ Illustration 8.9

Find $\log_{10} 29.31$.

$$29.31 = 2.931 \times 10^1$$
$$\log_{10} 29.31 = \log_{10} (2.931 \times 10^1)$$
$$= \log_{10} 2.931 + 1$$

Now, looking in Table III, Appendix B, we do not find $\log_{10} 2.931$. We do, however, find $\log_{10} 2.93$ and $\log_{10} 2.94$. For reasons that will be explained below, we do the following.

$$\log_{10} 2.93 = 0.4669$$
$$\log_{10} 2.931 = m$$
$$\log_{10} 2.94 = 0.4683$$

Since 2.931 is $\frac{1}{10}$ of the way from 2.93 to 2.94, we will approximate $m$ by a number that is one-tenth of the way from 0.4669 to 0.4683. The difference between 0.4683 and 0.4669 is

$$0.4683 - 0.4669 = 0.0014$$

Concentrating on the last two digits we see that one-tenth of 14 is $\frac{1}{10} \cdot 14 \equiv 1.4$. Rounding 1.4 off, we get 1. So we use $0.4669 + 0.0001 = 0.4670$ as our approximation for $m$. Finally,

$$\log_{10} 29.31 = \log_{10} 2.931 + 1$$
$$= 0.4670 + 1$$
$$= 1.4670 \quad \blacktriangleleft$$

Let us do one more illustration before discussing the process.

◆ **Illustration 8.10**

Find $\log_{10} 4.524$.

$$\log_{10} 4.52 = 0.6542$$
$$\log_{10} 4.524 = m$$
$$\log_{10} 4.53 = 0.6551$$

Since 4.524 is four-tenths of the way from 4.52 to 4.53, we approximate $m$ with a number that is four-tenths of the way from 0.6542 to 0.6551.

$$0.6551 - 0.6542 = 0.0009$$
$$\tfrac{4}{10} \cdot 9 = 3.6$$

Rounding off to 4 we use $m = 0.6542 + 0.0004 = 0.6546$ as our approximation. That is, we say

$$\log_{10} 4.524 = 0.6546 \quad \blacktriangleleft$$

What is going on here? Since there is an infinite number of numbers between 1 and 10, no table of $\log_{10} x$ for $1 \le x < 10$ can be complete. Also, since the range of each log function is the entire set of real numbers, many of the values of $\log_{10} x$ are irrational numbers. Irrational numbers have infinite nonrepeating decimal representations, so that any finite decimal representation of these numbers will be an approximation. Because there are infinitely many $x$'s and because many $\log_{10} x$'s are irrational, it is inevitable that we will need to approximate $\log_{10} x$ in many cases.

The approximating process we used in Illustrations 8.9 and 8.10 above is called *linear interpolation*. In Fig. 8.4a you can see why the adjective *linear* is used. (In the circular region the graph is almost a straight line.) In our approximation of $\log_{10} 4.524$ we did the following

$$\log_{10} 4.524 = \log_{10} 4.52 + \left(\frac{4.524 - 4.52}{4.53 - 4.52}\right)(\log_{10} 4.53 - \log_{10} 4.52)$$

(You should check to see that you agree that this is what we did.) Rewriting this expression we get

$$\frac{\log_{10} 4.524 - \log_{10} 4.52}{4.524 - 4.52} = \frac{\log_{10} 4.53 - \log_{10} 4.52}{4.53 - 4.52}$$

This expression states algebraically that the slope of the line from $P$ to $R$ in Fig. 8.4 is the same as the slope of the line from $P$ to $Q$. In other words, in choosing $\log_{10} 4.524$ as $0.6546$, we are treating the log function as though its graph between $P$ and $Q$ were the straight line from $P$ to $Q$. The error in this approximation is represented by the vertical distance between $R$ and the graph.

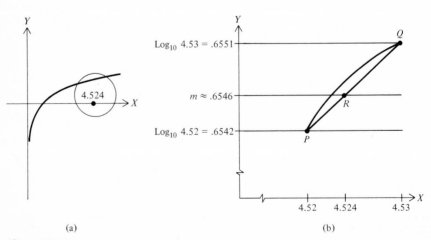

(a)                                                        (b)

**Figure 8.4**

The process of finding logarithms can be reversed. That is, given $\log_{10} x$, you can use Table III, Appendix B to find x.

### Illustration 8.11

If $\log_{10} x = 0.1271$, find x.

Looking in the table we see that $0.1271 = \log_{10} 1.34$. So we conclude that $x = 1.34$. ◀

### Illustration 8.12

If $\log_{10} x = 2.6551$, find x.

All the logarithms in the table are between 0 and 1 (i.e., between $\log_{10} 1$ and $\log_{10} 10$). So we look up 0.6551 and see that $0.6551 = \log 4.52$. We know that $2 = \log_{10} 10^2$. So

$$\begin{aligned}
\log_{10} x &= 2.6551 \\
&= 2 + 0.6551 \\
&= \log_{10} 10^2 + \log_{10} 4.52 \\
&= \log_{10} (4.52 \times 10^2) \\
&= \log_{10} 452
\end{aligned}$$

From this we conclude that $x = 452$. ◀

### Illustration 8.13

If $\log_{10} x = 0.6744$, find x.

This time you cannot find your number in the table. The number 0.6744 is not there. You can, however, find numbers on both sides of 0.6744.

$$\begin{aligned}
\log_{10} 4.72 &= 0.6739 \\
\log_{10} x &= 0.6744 \\
\log_{10} 4.73 &= 0.6749
\end{aligned}$$

Again we use linear interpolation. That is, we will choose x between 4.72 and 4.73 proportionally to the position of 0.6744 between 0.6739 and 0.6749.

$$\begin{aligned}
0.6739 - 0.6744 &= 0.0005 \\
0.6739 - 0.6749 &= 0.0010 \\
\frac{0.0005}{0.0010} &= 0.5
\end{aligned}$$

So we choose $x = 4.725$ to be halfway between 4.72 and 4.73. ◀

♦ **Illustration 8.14**

If $\log_{10} x = 0.7326$, find x.

$$\left.\begin{array}{l}\log_{10} 5.40 = 0.7324 \\ \log_{10} x = 0.7326 \\ \log_{10} 5.41 = 0.7332\end{array}\right\} {\scriptstyle 2} \Big\} {\scriptstyle 8}$$

Since $\frac{2}{8} = 0.25$ is rounded off to 0.3, we choose $x = 5.403$ which is three-tenths of the way from 5.40 to 5.41.  ◄

To this point we have only showed you how to find approximations to $\log_b$ for those cases where $b = 10$.  Fortunately there is a formula which will enable you to find $\log_b$ for any positive $b$ except 1 in terms of $\log_{10}$.

---

**THEOREM 8.4**
$$\log_b x = \frac{\log_{10} x}{\log_{10} b} \qquad b \neq 1, b > 0$$

**PROOF**
Let $y = \log_b x$.  Then $b^y = x$.  (Recall that the log and the exponential are inverses.)  So

$$\log_{10} b^y = \log_{10} x$$

and
$$y \log_{10} b = \log_{10} x$$

Thus
$$y = \frac{\log_{10} x}{\log_{10} b}$$

That is,
$$\log_b x = \frac{\log_{10} x}{\log_{10} b}$$

---

We will give an illustration of the use of this theorem before the problems for this section.

♦ **Illustration 8.15**

Find $\log_{2.72} 4.35$.
According to the theorem,

$$\log_{2.72} 4.35 = \frac{\log_{10} 4.35}{\log_{10} 2.72}$$

Looking at Table III, Appendix B, we get

$$\log_{10} 4.35 = 0.6385$$

and $\qquad\qquad\qquad \log_{10} 2.72 = 0.4346$

So $\log_{2.72} 4.35 = \dfrac{0.6385}{0.4346} = 1.4692$  ◀

## PROBLEMS 8.3

**1** Find each of the following.

    **a** $\log_{10} 3.51$     **b** $\log_{10} 952$

    **c** $\log_{10} 0.00175$     **d** $\log_{10} 45.39$

**2** Find each of the following.

    **a** $\log_{10} 7.32$     **b** $\log_{10} 27.6$     **c** $\log_{10} 397$     **d** $\log_{10} 0.02932$

**3** Solve each of the following for x.

    **a** $\log_{10} x = 0.6064$     **b** $\log_{10} x = 0.9300$

    **c** $\log_{10} x = 2 + 0.9059$     **d** $\log_{10} x = -3 + 0.4031$

**4** Solve each of the following for x.

    **a** $\log_{10} x = 0.7356$     **b** $\log_{10} x = 0.7312$

    **c** $\log_{10} x = -2 + 0.6713$     **d** $\log_{10} x = -0.3124$

**5** Find each of the following.

    **a** $\log_3 47$     **b** $\log_{7.321} 0.001235$

**6**   **a** $\log_{3.14} 273.4$     **b** $\log_{2.718} 51.32$

**7** Solve each of the following for x.

    **a** $\log_3 x = 2.3175$     **b** $\log_{25} x = 0.3012 - 2$

**8** Solve each of the following for x.

    **a** $\log_{2.72} x = 0.7213 - 3$     **b** $\log_{12} x = 1.4025$

**9** Draw a figure like Fig. 8.4 to illustrate the linear interpolation involved in computing $\log_{10} 2.436$.

## 8.4   APPLICATIONS OF LOGARITHMS

In this section you will study some of the uses to which logarithms can be put. In most cases we will use numbers that were used in Sec. 8.3 in order to minimize the computational complexity of the illustrations.

    Logarithms have long been used as a tool for making complex computations, especially those involving exponents, multiplication, and division. It is important to remember that most computations using logarithms are approximations and that many of the equality signs in these computations should really be thought of as approximate equality signs.

▶ **Illustration 8.16**

Compute $\dfrac{956 \cdot 168}{672}$

Let $N = \dfrac{956 \cdot 168}{672}$

Using Theorems 8.1 and 8.2 we get

$$\log_{10} N = \log_{10} 956 + \log_{10} 168 - \log_{10} 672$$

From Table III, Appendix B, we read that

$$\log_{10} 956 = 2 + \log_{10} 9.56 = 2 + 0.9805$$
$$\log_{10} 168 = 2 + \log_{10} 1.68 = 2 + 0.2253$$
$$\log_{10} 672 = 2 + \log_{10} 6.72 = 2 + 0.8274$$

So $\qquad \log_{10} N = (2 + 0.9805) + (2 + 0.2253) - (2 + 0.8274)$
$$= 2 + 0.3784$$

Looking again at the table we see that

$$0.3784 = \log_{10} 2.39$$

So $\qquad \log_{10} N = 2 + \log_{10} 2.39$
$$= \log_{10} 10^2 + \log_{10} 2.39$$
$$= \log_{10} (2.39 \times 10^2)$$
$$= \log_{10} 239$$

Thus $\qquad N = 239$ ◀

▶ **Illustration 8.17**

Compute $\sqrt[3]{\dfrac{29.31}{1.34 \cdot 327}}$

Let $N = \sqrt[3]{\dfrac{29.31}{1.34 \cdot 327}}$

$$\log_{10} N = \log_{10} \left( \frac{29.31}{1.34 \cdot 327} \right)^{1/3}$$
$$= \frac{1}{3} \log_{10} \frac{29.31}{1.34 \cdot 327} \qquad \text{Theorem 8.3}$$
$$= \tfrac{1}{3}(\log_{10} 29.31 - \log_{10} 1.34 - \log_{10} 327)$$

$$\log_{10} 29.31 = 1 + \log_{10} 2.931$$
$$= 1 + 0.4670 \qquad \text{by using linear interpolation on Table III, Appendix B}$$

$$\log_{10} 1.34 = 0.1271$$
$$\log_{10} 327 = 2 + \log 3.27 = 2 + 0.5145$$

Thus $\quad \log_{10} N = \frac{1}{3}[1 + 0.4670 - 0.1271 - (2 + 0.5145)]$
$$= \frac{1}{3}[-1 - 0.1746] = -0.3915$$

We have a problem. The numbers in the table are all positive (since they are logs of numbers between 1 and 10). Yet $\log_{10} N$ is negative. (This tells us that $N$ is between 0 and 1. Think of the graph.)

We perform a trick. We rewrite $-0.3915$ as $-1 + 1 - 0.3915$ (we just added and subtracted 1). Then, $-1 + 1 - 0.3915 = -1 + 0.6085$ ($1 - 0.3915 = 0.6085$). Interpolating the table we find that $0.6085 = \log_{10} 4.06$. So

$$\log_{10} N = -1 + \log_{10} 4.06$$
$$= \log_{10} 10^{-1} + \log_{10} 4.06$$
$$= \log_{10} (4.06 \times 10^{-1})$$
$$= \log_{10} 0.406$$

Thus $\qquad\qquad\qquad N = 0.406$ ◀

The computations above may seem messy, but the steps are fairly simple if you keep them straight.

To compute $N$,

**1** Use Theorems 8.1, 8.2, and 8.3 to express $\log_{10} N$ as a multiple of sums and differences of logs.
**2** Compute the needed logs using Table III, Appendix B and interpolating where needed.
**3** Combine the logs as dictated by step 1 to get $\log_{10} N$.
**4** Express $\log_{10} N$ as an integer plus a positive number between 0 and 1.
**5** Compute $N$ using the tables and interpolation as needed.

The fact that logs and exponentials are inverses is directly useful in solving equations which involve exponentials.

◀ **Illustration 8.18**

How many years are required for a checking account to double at an annual interest rate of 5 percent? In Chap. 6 we found that the balance $B_n$ in the account after $n$ years will be $B_n = B_0(1.05)^n$, where $B_0$ is the initial balance. We need to solve the following equation for $n$.

$$B_0(1.05)^n = 2B_0$$

First, we see that the equation can be simplified by dividing out $B_0$.

$$(1.05)^n = 2$$

Using logs we proceed as follows.

$$\log_{10}(1.05)^n = \log_{10} 2$$
$$n \log_{10} 1.05 = \log_{10} 2$$
$$n = \frac{\log_{10} 2}{\log_{10} 1.05}$$
$$\log_{10} 2 = 0.3010$$
$$\log_{10} 1.05 = 0.0212$$

So
$$n = \frac{0.3010}{0.0212} = 14.198$$

Thus we see that after the fifteenth year a 5 percent annual rate savings account will have doubled. ◀

▶ **Illustration 8.19**

Radioactive substances decay with time. For a certain substance it was found that the amount of the substance present after $t$ years is given by the formula $A(t) = A_0 e^{-2t}$, where $A_0$ is the initial amount of the substance and $e$ is an irrational number whose first four digits are 2.718. Find how long it would take for half of the substance to decay. To do this we need to solve the following equation for $t$.

$$A_0 e^{-2t} = \tfrac{1}{2}A_0$$

Simplifying, we get

$$e^{-2t} = \tfrac{1}{2}$$

So
$$\log_{10} e^{-2t} = \log_{10} \tfrac{1}{2}$$
$$-2t \log_{10} e = \log_{10}\tfrac{1}{2}$$
$$t = -\frac{1}{2}\left(\frac{\log_{10}\tfrac{1}{2}}{\log_{10} e}\right)$$
$$\log_{10}\tfrac{1}{2} = \log_{10} 0.5 = 0.6990 - 1$$
$$\log_{10} e = \log_{10} 2.718 = 0.4343$$

Thus
$$t = -\frac{1}{2}\left(\frac{-1 + 0.6990}{0.4343}\right) = 0.3465$$

So we see that it takes a little more than one-third of a year for half the radioactive material to decay. ◀

## PROBLEMS 8.4

**1** Compute N in each case.

**a** $N = \dfrac{0.4932 \times 653.7}{0.07213 \times 8456}$

**b** $N = \sqrt{\dfrac{48.27 \times (0.3032)^2}{71.36}}$

**c** $N = \sqrt{0.532}\ \sqrt[3]{0.829}$

**2** Compute N in each case.

**a** $N = \dfrac{919}{83.3 \times 3.77}$

**b** $N = \dfrac{\sqrt[7]{7.77}}{\sqrt[4]{0.444}}$

**c** $N = \left[\dfrac{(28.44)^2\ \sqrt{0.5828}}{345.6}\right]^{1/3}$

**3** **a** If $4(1.72)^x = 297$, find x. **b** If $3.47^{1/x} = 0.001$, find x.

**4** **a** If $(1.0752)^x = 0.327$, find x. **b** If $8^{-3x} = 2.375$, find x.

**5** In how many years will a savings account double if it pays an annual interest rate of 6.5 percent?

**6** In how many years will a savings account triple if it pays a semiannual interest rate of 5.75 percent?

**7** A certain population of bacteria grows very rapidly, so that the number of bacteria in the population at time $t$ minutes is given by the formula $N(t) = 7.31e^{3t}$ (where $e$ can be approximated by 2.718). How long will it take for the population to reach 5 million in number?

**8** A radioactive material is decaying so that the amount present at time $t$ is given by $A(t) = 3e^{0.0032t}$. How long is the half-life of the material? That is, how long will it take for $A(t) = \frac{1}{2}A(0)$?

# TRIGONOMETRIC FUNCTIONS

## 9.1 TRIGONOMETRY OF REAL NUMBERS

The subject of trigonometry, dealing with angles and triangles both in the plane and on the sphere, was well established before the time of Christ. Its principal uses were in the fields of astronomy, navigation, land surveying, and construction. This aspect of the trigonometry of angles has faded in importance with the coming of higher mathematics, electronics, computers, etc. Today the emphasis is on the trigonometry of real numbers, and we believe you will have less difficulty in grasping the basic notions if we present the subject first without using the notion of angle, reserving the trigonometry of angles as a matter of secondary consideration.

## 9.2 ARC LENGTH

The concept of the length of an arc of a circle is mathematically very complicated. Intuitively it seems as if we could measure the length of an arc by wrapping a string around the circle and then measuring the length of the string. This is about the best we can do without a lot of higher mathematics.

The circumference $C$ of a circle of radius $r$ is given by

$$C = 2\pi r$$

where $\pi$ is irrational and therefore in decimal form does not repeat. The following rational number is an approximation for $\pi$:

$$\pi = 3.141592653589793$$

In 1962, an electronic digital computer, making 100,000 calculations each second (on numbers containing 20 decimal places) computed $\pi$ to 100,000 decimal places. While making 31,380,000,000 calculations, it took the computer a mere 8 hours and 43 minutes.

The circumference of a circle of unit radius is $2\pi$, or about 6.28. The circumference of the circle $x^2 + y^2 = 16$ is $2\pi \times 4 = 25.12$. The radius of a circle with circumference 628 centimeters is about 100 centimeters.

Consider a unit circle (radius = 1) with center placed at the origin of a rectangular coordinate system. The pythagorean theorem says that the points $(x, y)$ on this circle satisfy the equation $x^2 + y^2 = 1$ (Fig. 9.1). Keep a sharp eye on Fig. 9.1 as you read the rest of this section.

Think of the line tangent to the unit circle at $P(1, 0)$ as a number line and mark off the scale units. Now wind the positive (upper) half of the number line (as a string) around the circle counterclockwise from $P$ to $Q$. Let $S$ be the length of the arc $PQ$. Unwind the string back into the number line. Point $Q$, on the string, will fall on point $Q'$ on the real number line, and $S$, the length of arc $PQ$, will be the straight-line length $PQ'$ on the real number line. We shall speak indifferently of *length* of arc $S$ and *arc* $S$. Similarly, wind counterclockwise around to $R$ and let $S_1$ be the length of arc $PQR$. Unwind. Point $R$ will fall on $R'$, and $S_1$, the length of arc $PQR$, will be $PR'$ or the real number $R'$. We could wind all the way around to $P$, getting the arc $C$ (circumference), and an unwinding would place this arc length at a distance $2\pi$ up on the real number line. We could keep on winding around and around. Unwinding would yield more and more arcs and associated numbers on the positive half of the number line.

Winding the negative half of the number line clockwise we get arcs (we will call them negative arcs) associated with negative real numbers. So hereafter we must specify the direction of the winding and hence we speak of *directed* arcs: Those counterclockwise are positive, and those clockwise are negative. Clearly associated with each arc there is one and only one real number, and associated with each

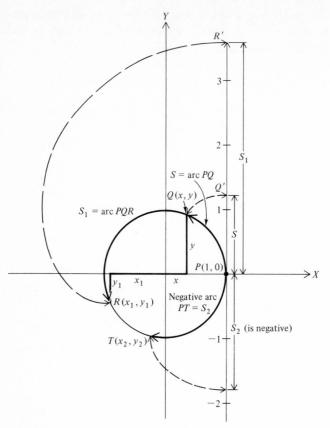

**Figure 9.1**

real number there is one and only one arc. Of course there are many arcs with the same terminal position, such as Q in Fig. 9.1. There is the arc S as exhibited, and there are S plus one revolution, S plus two revolutions, etc. But there is only one arc, namely $S = PQ$, associated with the real number $Q'$. We say there is a one-to-one correspondence between arcs (arc lengths) and real numbers.

## PROBLEMS 9.2

*In Probs. 1 to 10, on a unit circle, draw the arc for the given value of S.*

| | | | |
|---|---|---|---|
| **1** $S = 1$ | **2** $S = 2.5$ | **3** $S = 3.14$ | **4** $S = 6.28$ |
| **5** $S = 10$ | **6** $S = -4$ | **7** $S = -3$ | **8** $S = -3.14$ |
| **9** $S = -12$ | **10** $S = -12.56$ | | |

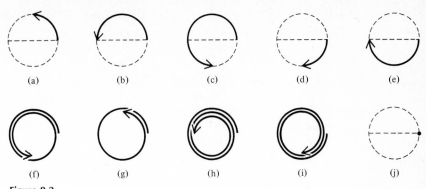

(a)         (b)         (c)         (d)         (e)

(f)         (g)         (h)         (i)         (j)

**Figure 9.2**

**11**   Estimate the length of the given directed arc in Fig. 9.2a–j by mentally unwinding it.

## 9.3   SINE, COSINE, AND TANGENT

We are now in a position to define three trigonometric functions called *sine*, *cosine*, and *tangent*. For greater clarity we have stripped down Fig. 9.1 to the bare essentials. These appear in Fig. 9.3.

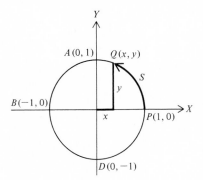

**Figure 9.3**

## DEFINITIONS 9.1

Given a real number S, representing the length of a directed arc PQ (which may include any number of windings all the way around the circle), let (x, y) be the coordinates of the unique endpoint. Then *sin S* and *cos S* are defined to be

$$\sin S = y$$
$$\cos S = x$$

These are read "The sine of the real number S is the real number y" and "The cosine of the real number S is the real number x." The sine and cosine are therefore functions of the real number S. The domain is the set of real numbers. From Fig. 9.3 (and keeping in mind the fact that the radius of the circle is 1), it should be clear that y cannot be greater than 1 and cannot be less than −1. The range of sine is therefore all real numbers between −1 and 1 inclusive. A similar argument applies to x, and so the range of cosine is likewise all real numbers between −1 and 1 inclusive.

## DEFINITION 9.2

The third trigonometric function, *tangent*, is defined to be

$$\tan S = \frac{\sin S}{\cos S} \qquad \cos S \neq 0$$

The domain of the tangent function is all real numbers S, except those for which $\cos S = 0$. ◀

Whenever the endpoint of the directed arc S lies on the Y axis, it follows that x is 0. For such an arc S the value of the sine is 1, or −1, the value of the cosine is 0, and the tangent does not exist. To find the range of the tangent we can argue as follows. When $S = 0$, Q coincides with P and $y/x = 0/1 = 0$. When S is positive and small, Q is near P and [y (small)]/[x (near 1)] is a small positive number. For the S in Fig. 9.3, $y/x > 1$ surely. To our eyes $y/x$ is about 2.5. As Q approaches the Y axis, S approaches the number $C/4 = 2\pi/4 = \pi/2 = 1.57$, y approaches 1, and x approaches 0. And [y (near 1)]/[x (near 0)] is a very large number. Therefore the range of tangent in the first quadrant is 0 and all positive reals. A similar argument applies for the second quadrant, but there x is negative and hence $\tan S \leq 0$. The full range of tan S is therefore all real numbers.

Table II, Appendix B gives values of sine, cosine, and tangent (and cotangent, abbreviated cot, where by definition $\cot = 1/\tan$, $\tan \neq 0$) for some real numbers in the interval 0 to $2\pi$.

Finally note that, since $\sin S = y$ and $\cos S = x$ and, by the pythagorean theorem, $x^2 + y^2 = 1$, it follows that, for all arcs S,

$$(\sin S)^2 + (\cos S)^2 = 1 \tag{1}$$

This is called an *identity*, since it is true for all arcs. We usually write it $\sin^2 S + \cos^2 S = 1$. (Do not confuse $\sin^2 S$ with $\sin S^2$.) We shall see other identities; this is one of the most important in trigonometry.

## PROBLEMS 9.3

In Probs. 1 to 16, use Table II, Appendix B, to find the sine, cosine, and tangent of the indicated real number.

| | | | | | | | |
|---|---|---|---|---|---|---|---|
| **1** | 0.75 | **2** | 0.98 | **3** | 1.61 | **4** | 1.19 |
| **5** | −1.79 | **6** | −1.47 | **7** | 0.11 | **8** | 0.24 |
| **9** | −1.82 | **10** | −1.01 | **11** | −1.57 | **12** | −2.00 |
| **13** | 0.24 | **14** | 1.60 | **15** | 1.98 | **16** | 0.75 |

## 9.4    PERIODICITY

Figure 9.4 shows an arc $PA = S$. Unwound, $A$ falls on $A'$, and $S = PA'$ in length. For the real number $S$, $y = \sin S$, and $x = \cos S$. For the arc pictured, $y$ is positive and $x$ is negative. Now consider the real number $S + 2\pi$. The arc for this number is arc $S$ plus one complete revolution (winding around the circumference of the unit circle), that is, $S + 2\pi$. Obviously, $y = \sin (S + 2\pi)$ and $x = \cos (S + 2\pi)$. Un-

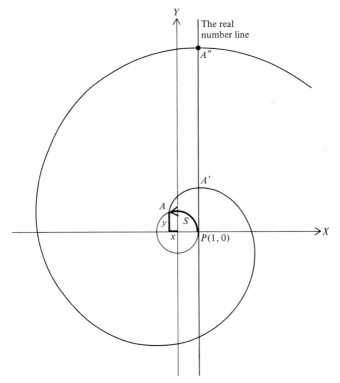

**Figure 9.4**

wound this runs from $P$ to $A''$. Since $PA' = S$, therefore $A'A'' = 2\pi$. For any arc $S$ it is true that $\sin S = \sin(S + 2\pi)$. Similarly, $\cos S = \cos(S + 2\pi)$. Moreover any number of complete revolutions (complete circular windings) will give similar results, that is, $\sin S = \sin(S + 2n\pi)$ and $\cos S = \cos(S + 2n\pi)$, where $n$ is an integer. We say that sin and cos are periodic functions of period $2\pi$.

It is helpful to know how the trigonometric functions behave in the various quadrants. In Quadrants I and II, $y \geq 0$, while in III and IV, $y \leq 0$. In Quadrants I and IV, $x \geq 0$, and in II and III, $x \leq 0$. Displayed in Table 9.1, they help us in seeing at a glance the signs of sin, cos, and tan (omitting zero values).

**Table 9.1**

| Function | Quadrant | | | |
|----------|:---:|:---:|:---:|:---:|
|          | I | II | III | IV |
| sin | + | + | − | − |
| cos | + | − | − | + |
| tan | + | − | + | − |

The signs of the tan function need a little explanation. In Quadrant I, both sine and cosine are positive, and so the ratio sine/cosine (= tangent) is positive. In Quadrant II, sine is positive and cosine is negative, and therefore tangent is negative. In Quadrant III, sine and cosine are both negative, but their ratio is positive and hence tangent is positive. For Quadrant IV, sine is negative, cosine is positive, and tangent is negative (Fig. 9.5).

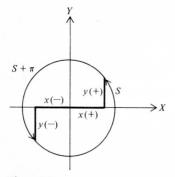

**Figure 9.5**

The tangent function is also periodic, the period being $\pi$. In Fig. 9.5 it is seen that

$$\tan S = \frac{y(+)}{x(+)} = \frac{y(-)}{x(-)} = \tan (S + \pi)$$

### DEFINITION 9.3

A function $f$ such that $f(x) = f(x + p)$ for some positive $p$ and all $x$ is said to be a *periodic function*. The least number $p$ for which this is true is called the *period* of the function. The period for sin and cos is $2\pi$. The period for tan is $\pi$. ◀

## PROBLEMS 9.4

*In Probs. 1 to 6 fill in the values where they exist.*

| | arc S | 0 | $\pi/2$ | $\pi$ | $3\pi/2$ | $2\pi$ | $5\pi/2$ | $3\pi$ | $7\pi/2$ | $4\pi$ |
|---|---|---|---|---|---|---|---|---|---|---|
| 1 | sin S | | | | | | | | | |
| 2 | cos S | | | | | | | | | |
| 3 | tan S | | | | | | | | | |

| | arc S | 0 | $-\pi/2$ | $-\pi$ | $-3\pi/2$ | $-2\pi$ |
|---|---|---|---|---|---|---|
| 4 | sin S | | | | | |
| 5 | cos S | | | | | |
| 6 | tan S | | | | | |

## 9.5   SPECIAL NUMBERS

In plane geometry it was proved that the side of a hexagon inscribed in a circle of radius $r$ is $r$ in length (Fig. 9.6, where $r = 1$). For an arc of $\pi/6$ we have $OQ = 1$, $PQ = \frac{1}{2}$, and by the pythagorean theorem $(x^2 + y^2 = 1)$ it follows that

$$x^2 + (\tfrac{1}{2})^2 = 1^2$$

or
$$x^2 + \tfrac{1}{4} = 1$$
$$x^2 = \tfrac{3}{4}$$
$$x = \pm\tfrac{1}{2}\sqrt{3}$$

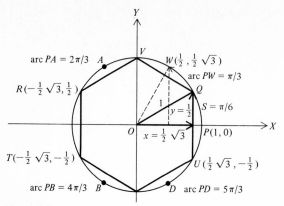

**Figure 9.6**

Since for arc S, x is positive, we can write down the values of sin, cos, and tan of arc $S = \pi/6$.

$$\sin\frac{\pi}{6} = \frac{1}{2}$$

$$\cos\frac{\pi}{6} = \frac{1}{2}\sqrt{3}$$

$$\tan\frac{\pi}{6} = \frac{1}{\sqrt{3}} = \frac{1}{3}\sqrt{3}$$

Arc $PQR = 5\pi/6$, arc $PQT = 7\pi/6$, and arc $PQU = 11\pi/6$; and from the coordinates we can write down sin, cos, and tan of any arc with endpoint at Q, R, T, or U.

If we bisect side QV, then W has coordinates $(\frac{1}{2}, \frac{1}{2}\sqrt{3})$. The x and the y for W are the y and x, respectively, for Q. The arc PW is twice the arc PQ, so arc $PQ = \pi/3$. In the second, third, and fourth quadrants the corresponding arcs terminate at $A(-\frac{1}{2}, \frac{1}{2}\sqrt{3})$, $B(-\frac{1}{2}, -\frac{1}{2}\sqrt{3})$, and $D(-\frac{1}{2}, -\frac{1}{2}\sqrt{3})$, and their coordinates determine sin, cos, and tan.

It is even simpler to determine the sin, cos, and tan of arcs terminating at $\pi/4$, $3\pi/4$, $5\pi/4$, and $7\pi/4$ (Fig. 9.7). Let the X axis bisect the opposite sides of the inscribed square. Then the Y axis bisects the other two sides. Therefore, for arc PQ, x and y are equal. By the pythagorean theorem, $x^2 + x^2 = 1$ or $2x^2 = 1$, $x^2 = \frac{1}{2}$, $x = \pm 1/\sqrt{2} = \frac{1}{2}\sqrt{2}$. From the coordinates of Q, R, T, and U, the values of the sin, cos, and tan can be read off.

## PROBLEMS 9.5

In Probs. 1 to 6 make use of Figs. 9.6 and 9.7 to fill in the values of sin, cos, and tan.

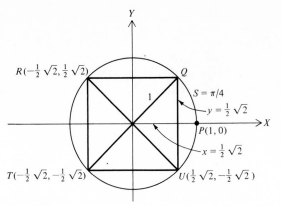

**Figure 9.7**

**1**

| arc S | π/6 | 5π/6 | 7π/6 | 11π/6 |
|-------|-----|------|------|-------|
| sin S |     |      |      |       |
| cos S |     |      |      |       |
| tan S |     |      |      |       |

**2**

| arc S | π/3 | 2π/3 | 4π/3 | 5π/3 |
|-------|-----|------|------|------|
| sin S |     |      |      |      |
| cos S |     |      |      |      |
| tan S |     |      |      |      |

**3**

| arc S | π/4 | 3π/4 | 5π/4 | 7π/4 |
|-------|-----|------|------|------|
| sin S |     |      |      |      |
| cos S |     |      |      |      |
| tan S |     |      |      |      |

|   | arc S | 0 | π/6 | π/4 | π/3 | π/2 | 2π/3 | 3π/4 | 5π/6 | π |
|---|-------|---|-----|-----|-----|-----|------|------|------|---|
| **4** | sin S |   |     |     |     |     |      |      |      |   |
| **5** | cos S |   |     |     |     |     |      |      |      |   |
| **6** | tan S |   |     |     |     |     |      |      |      |   |

## 9.6    GRAPHS

By making use of the trigonometric values associated with the special numbers only, we can draw reasonably accurate graphs of sin, cos, and tan. All the needed values of x and y are contained in Figs. 9.6, 9.7, and 9.8. Figure 9.8 yields the following ordered pairs, $(S, y)$, for $y = \sin S$:

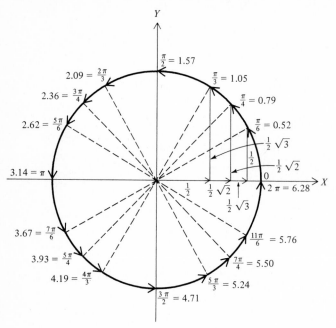

$\frac{1}{2}\sqrt{2} = 0.71$ (rounded); $\frac{1}{2}\sqrt{3} = 0.87$ (rounded)

**Figure 9.8**

$(0, 0)$, $(0.52, 0.50)$, $(0.79, 0.71)$, $(1.05, 0.87)$, $(1.57, 1.00)$,
$(2.09, 0.87)$, $(2.36, 0.71)$, $(2.62, 0.50)$, $(3.14, 0)$, $(3.67, -0.50)$,
$(3.93, -0.71)$, $(4.19, -0.87)$, $(4.71, -1.00)$, $(5.24, -0.87)$,
$(5.50, -0.71)$, $(5.76, -0.50)$, $(6.28, 0)$

These are plotted in Fig. 9.9.

Note that the portion of the curve in the first quadrant is essentially triplicated in the other three quadrants. The curve in the second quadrant is the mirror image of that in the first, the "mirror" being the line whose equation is $S = \pi/2$. Also, if the graph in the first *and* second quadrants is pushed (*translated* is the mathematical term) to

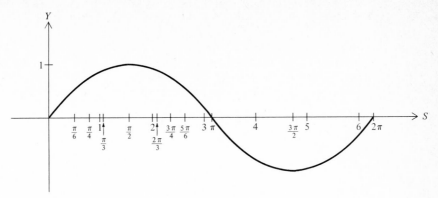

**Figure 9.9**

the right until it spans the third and fourth quadrants, then *its* mirror image about the S axis is the rest of the curve.

Figure 9.8 also yields the ordered pairs for x = cos S. You will have to perform a few divisions to get tan S = sin S/cos S.

So far the notations using S, x, and y have been desirable. But now look at the ordered pairs for the sine function. We might just as well call the first member of each pair x and the second y. Thus we would be back with our standard notation and we would plot these points with respect to coordinate axes X and Y. For the cos function, up to now given by x = cos S, we would once again write x and y; thus y = cos x. And likewise y = tan x.

Figure 9.10 shows the graphs of y = sin x, y = cos x and y = sin x + cos x. Figure 9.11 shows y = sin x, y = cos x, and y = tan x drawn to the same scale on the two axes. It also shows their periodic character.

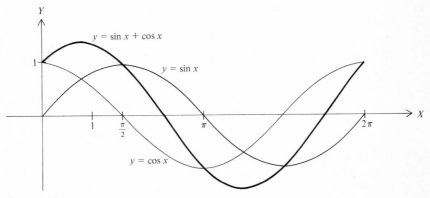

**Figure 9.10**

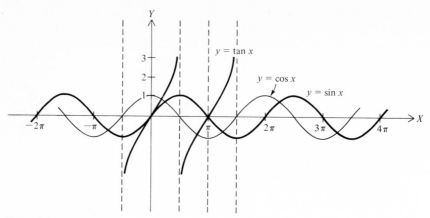

**Figure 9.11**

The maximum value of $y = \sin x$ is 1 and occurs where $x = \pi/2 + 2n\pi$, $n$ an integer. The maximum value is called the *amplitude* of the sine wave, and a similar remark applies to $y = \cos x$ whose amplitude is 1 and occurs where $x = 2n\pi$, $n$ an integer. For $y = A \sin x$ *the amplitude is* $|A|$. This should be obvious since each ordinate on $y = A \sin x$ is $A$ times the corresponding ordinate on $y = \sin x$. The amplitude $A$ does not affect the period.

Let us compare the curves whose equations are $y = \sin x$ and $y = \sin 2x$. Examine once again Fig. 9.9. The curve for $y = \sin x$ rises from 0 to 1 in the first quadrant, falls back to 0 in the second, continues to decrease (to $-1$) in the third, and increases to 0 in the fourth. The curve for $y = \sin 2x$ rises from 0 to 1 as x increases from 0 to $\pi/4$, because $2x = \pi/2$ when $x = \pi/4$. As x increases from $\pi/4$ to $\pi/2$, 2x increases from $\pi/2$ to $\pi$, and the curve therefore falls from 1 to 0. It should be clear that $y = \sin 2x$ finishes a full period pattern in the interval $0 \le x \le \pi$, since in that interval 2x covers the interval $0 \le 2x \le 2\pi$. In other words the "sine wave" $y = \sin 2x$ oscillates twice as fast as the sine wave $y = \sin x$. The period for $y = \sin 2x$ is $\pi$, one-half that of $y = \sin x$, which is $2\pi$ (Fig. 9.12).

Finally we should like to compare the graphs of $y = \sin x$ and $y = \sin (x + \pi/6)$. Now for $y = \sin x$, $y = 0$ when $x = 0$. For $y = \sin (x + \pi/6)$, $y = 0$ when $x + \pi/6 = 0$, namely when $x = -\pi/6$. So $y = \sin (x + \pi/6)$ is ahead of $y = \sin x$, starting to rise at $x = -\pi/6$. Otherwise it is a duplicate copy of $y = \sin x$ (Fig. 9.13). The sine wave is said to have undergone a *phase shift*. The phase shift for $y = \sin (x + \pi/2)$ is $-\pi/2$, and the resulting curve coincides with the cosine curve $y = \cos x$. For all x, $\sin (x + \pi/2) = \cos x$ identically.

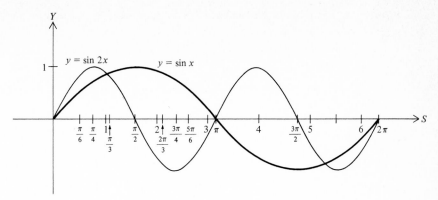

**Figure 9.12**

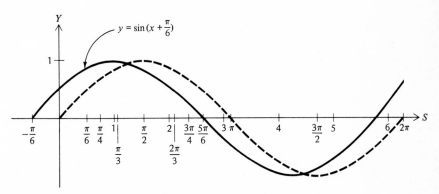

**Figure 9.13**

## PROBLEMS 9.6

In Probs. 1 to 6 sketch each of the two curves on the same axes and to the same scale, in the interval $0 \le x \le 2\pi$ for sine and cosine, and in the interval $0 \le x \le \pi$ for tangent.

**1** $y = \sin x$ and $y = 2 \sin x$      **2** $y = \sin x$ and $y = \frac{1}{2} \sin x$

**3** $y = \cos x$ and $y = 2 \cos x$      **4** $y = \cos x$ and $y = \frac{1}{2} \cos x$

**5** $y = \tan x$ and $y = 2 \tan x$      **6** $y = \tan x$ and $y = \frac{1}{2} \tan x$

In Probs. 7 to 12 sketch each of the two graphs on the same axes and to the same scale.

**7** $y = \sin x$ and $y = \sin \frac{1}{2}x$      **8** $y = \cos x$ and $y = \cos 2x$

**9** $y = \cos x$ and $y = \cos \frac{1}{2}x$      **10** $y = \sin x$ and $y = \sin 3x$

**11** $y = \cos x$ and $y = \cos 3x$      **12** $y = \tan x$ and $y = \tan 2x$

In Probs. 13 to 18 sketch each of the two graphs on the same axes and to the same scale.

**13**  $y = \sin x$ and $y = \sin\left(x + \dfrac{\pi}{3}\right)$    **14**  $y = \cos x$ and $y = \cos\left(x + \dfrac{\pi}{3}\right)$

**15**  $y = \sin x$ and $y = \sin\left(x - \dfrac{\pi}{3}\right)$    **16**  $y = \cos x$ and $y = \cos\left(x - \dfrac{\pi}{3}\right)$

**17**  $y = \sin x$ and $y = \sin\left(x - \dfrac{\pi}{2}\right)$    **18**  $y = \cos x$ and $y = \cos\left(x - \dfrac{\pi}{2}\right)$

In Probs. 19 to 24 sketch each of the three graphs on the same axes and to the same scale.

**19**  $y = \sin x$, $y = 2 \sin x$, and $y = 2 \sin \frac{1}{2}x$
**20**  $y = \sin x$, $y = \frac{1}{2} \sin x$, and $y = \frac{1}{2} \sin 2x$
**21**  $y = \cos x$, $y = \cos 2x$, and $y = \frac{1}{2} \cos 2x$
**22**  $y = \cos x$, $y = \cos \frac{1}{2}x$, and $y = 2 \cos \frac{1}{2}x$
**23**  $y = \sin 2x$, $y = \sin\left(2x + \dfrac{\pi}{4}\right)$, and $y = 2 \sin\left(2x + \dfrac{\pi}{4}\right)$
**24**  $y = \cos 2x$, $y = \cos\left(2x - \dfrac{\pi}{4}\right)$, and $y = 2 \cos\left(2x - \dfrac{\pi}{4}\right)$

## 9.7    COSECANT, SECANT, AND COTANGENT

In both mathematical theory and its applications to the physical world, the sin, cos, and tan functions occur in reciprocal form as 1/sin, 1/cos, and 1/tan. We give other names to these.

### DEFINITIONS 9.4

By definition, cosecant = 1/sin, secant = 1/cos, and cotangent = 1/tan, and these are abbreviated to csc, sec, and cot, respectively.  ◀

Since we never divide by 0, csc does not exist when sin is 0, sec does not exist when cos is 0, and cot does not exist when tan is 0.

We make regular use of these six trigonometric functions because they simplify both our formulas and our thinking even though, once we know the value of any one of them for a given number x, the other five may be computed. Suppose that for a given x we know sin x. Then, from the basic definitions, we have, for $0 \le x \le \pi/2$,

$$\cos x = \sqrt{1 - \sin^2 x} \qquad \text{pythagorean theorem}$$

$$\tan x = \frac{\sin x}{\cos x} = \frac{\sin x}{\sqrt{1 - \sin^2 x}}$$

$$\csc x = \frac{1}{\sin x}$$

$$\sec x = \frac{1}{\cos x} = \frac{1}{\sqrt{1 - \sin^2 x}}$$

and
$$\cot x = \frac{\sqrt{1 - \sin^2 x}}{\sin x}$$

We have thus expressed the other five functions in terms of sin alone, but some of the formulas are inelegant to say the least. Tables exist for each of the six functions.

## PROBLEMS 9.7

**1** For a given number $x$, express the other five trigonometric functions in terms of $\cos x$.

*In Probs. 2 to 4, use the special values of sin x, cos x, and tan x and from these compute csc x = 1/sin x, sec x = 1/cos x, and cot x = 1/tan x, and sketch in the interval $0 \leq x \leq 2\pi$.*

**2** $y = \csc x$     **3** $y = \sec x$     **4** $y = \cot x$

## 9.8   ADDITION THEOREMS

Let us consider a function $f$ and its values $f(x_1)$ and $f(x_2)$, where $x_1$ and $x_2$ are to be thought of as any two $x$'s in the domain of definition such that $x_1 + x_2$ is also in the domain. The following general question arises: "What can we say about $f(x_1 + x_2)$ in terms of $f(x_1)$ and $f(x_2)$ separately?" Such a theorem is referred to as an *addition theorem for the function f.* Both classical and modern mathematics place emphasis on the discovery and use of such theorems. They are of special importance in trigonometry where we should like to know the answers to the following questions:

**a** Can we express $\sin(\theta \pm \phi)$ and $\cos(\theta \pm \phi)$ in terms of $\sin \theta$, $\sin \phi$, $\cos \theta$, and $\cos \phi$? (Letters from the Greek alphabet are often used in trigonometry.)

**b** If so, what are the formulas?

There are many derivations of these formulas, and they all have some degree of artificiality. This can be said about most of the theorems of plane geometry where certain construction lines — once they have been thought of and drawn in — aid in the analysis of the problem.

We want to find a formula for one of the four quantities $\sin(\theta \pm \phi)$, $\cos(\theta \pm \phi)$, and therefore we begin by drawing two unit circles as in Fig. 9.14. In Fig. 9.14a we have laid off the arcs $\theta$ and $\phi$ and indicated the resulting arc $\phi - \theta$ as the arc $QP$. In Fig. 9.14b we have laid off the arc $\phi - \theta$ as the arc $SR$. The coordinates of the points $P$, $Q$, $R$, and $S$ are indicated on the figure; their values follow from the definitions of sine and cosine.

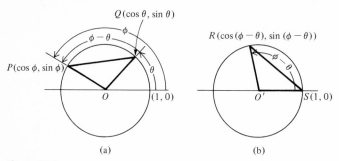

(a)                              (b)

**Figure 9.14**

The key to our result is the remark that the segments $PQ$ and $RS$ are equal, for they are corresponding sides of the congruent triangles $OPQ$ and $O'RS$. If we express this equality by the use of the distance formula (Sec. 4.5), we find

$$(PQ)^2 = (\cos\phi - \cos\theta)^2 + (\sin\phi - \sin\theta)^2$$
$$= 2 - 2(\cos\phi\cos\theta + \sin\phi\sin\theta)$$
$$(RS)^2 = [\cos(\phi - \theta) - 1]^2 + [\sin(\phi - \theta) - 0]^2$$
$$= 2 - 2\cos(\phi - \theta)$$

Equating these, we obtain

$$2 - 2\cos(\phi - \theta) = 2 - 2(\cos\phi\cos\theta + \sin\phi\sin\theta)$$

Therefore we have the formula

$$\cos(\phi - \theta) = \cos\phi\cos\theta + \sin\phi\sin\theta \qquad (1)$$

This is exactly the kind of formula we are seeking. It is an identity which expresses the cosine of the difference of two real numbers $\phi$ and $\theta$ in terms of the sines and cosines of $\phi$ and $\theta$ separately. You must memorize formula (1); it is one of the addition theorems for the trigonometric functions. We now derive three others, namely (4), (8), and (9) below.

Directly from the definitions of $\sin\theta$ and $\cos\theta$ it follows that

$$\sin(-\theta) = -\sin\theta \tag{2}$$

and
$$\cos(-\theta) = \cos\theta \tag{3}$$

In (1) put $\theta = -\alpha$. [We may do this since (1) is an identity.] We get $\cos(\phi + \alpha) = \cos\phi\cos(-\alpha) + \sin\phi\sin(-\alpha)$. Using (2) and (3), this simplifies to the second important addition theorem:

$$\cos(\phi + \alpha) = \cos\phi\cos\alpha - \sin\phi\sin\alpha \tag{4}$$

Next in (1) let us put $\theta = \pi/2$. We get

$$\cos\left(\phi - \frac{\pi}{2}\right) = \cos\phi\cos\frac{\pi}{2} + \sin\phi\sin\frac{\pi}{2}$$

or
$$\cos\left(\phi - \frac{\pi}{2}\right) = \sin\phi \tag{5}$$

If we put $\alpha = \phi - \pi/2$, or $\phi = \alpha + \pi/2$ in (5), we can write (5) in the form:

$$\cos\alpha = \sin\left(\alpha + \frac{\pi}{2}\right) \tag{6}$$

Similarly by putting $\theta = -\pi/2$ in (1), we obtain

$$\cos\left(\phi + \frac{\pi}{2}\right) = -\sin\phi \tag{7}$$

We are now ready to derive the addition theorem for $\sin(\phi - \theta)$. We use (5) to write

$$\sin(\phi - \theta) = \cos\left[(\phi - \theta) - \frac{\pi}{2}\right] = \cos\left[\phi - \left(\theta + \frac{\pi}{2}\right)\right]$$

Now apply (1) to the right-hand expression and obtain

$$\sin(\phi - \theta) = \cos\phi\cos\left(\theta + \frac{\pi}{2}\right) + \sin\phi\sin\left(\theta + \frac{\pi}{2}\right)$$

Using (6) and (7), we can simplify this to

$$\sin(\phi - \theta) = -\cos\phi\sin\theta + \sin\phi\cos\theta$$

or
$$\sin(\phi - \theta) = \sin\phi\cos\theta - \cos\phi\sin\theta \tag{8}$$

This is the desired result.

Finally, putting $\theta = -\alpha$ in (8), we obtain

$$\sin(\phi + \alpha) = \sin\phi\cos\alpha + \cos\phi\sin\alpha \tag{9}$$

We now collect (1), (4), (8), and (9) and write them in the form:

$$\sin(\phi \pm \theta) = \sin\phi\cos\theta \pm \cos\phi\sin\theta \tag{I}$$

$$\cos(\phi \pm \theta) = \cos\phi\cos\theta \mp \sin\phi\sin\theta \tag{II}$$

You should study this material until you understand it thoroughly. Be sure to note that (I) and (II) are identities which hold for all values of $\theta$ and $\phi$. They should be memorized.

**PROBLEMS 9.8**

**1** Making use of (I) and (II), show that

    **a**   $\tan(\phi \pm \theta) = \dfrac{\tan \phi \pm \tan \theta}{1 \mp \tan \phi \tan \theta}$

    **b**   $\sin 2\theta = 2 \sin \theta \cos \theta$

    **c**   $\cos 2\theta = \cos^2 \theta - \sin^2 \theta$

    **c′**  $\cos 2\theta = 2 \cos^2 \theta - 1$    Also use Eq. (1) in Sec. 9.3

    **c″**  $\cos 2\theta = 1 - 2 \sin^2 \theta$    Also use Eq. (1) in Sec. 9.3

We define a first, second, third, or fourth quadrantal arc $\theta$ as one such that when $\theta$ is laid off from $(1, 0)$ its endpoint lies in the first, second, third, or fourth quadrant, respectively.

**2** Use parts (c″) and (c′) of Prob. 1 to show that

    **a**  $\sin \dfrac{\theta}{2} = \begin{cases} \sqrt{\dfrac{1 - \cos \theta}{2}} & \dfrac{\theta}{2} \text{ a first or second quadrantal arc} \\[3ex] -\sqrt{\dfrac{1 - \cos \theta}{2}} & \dfrac{\theta}{2} \text{ a third or fourth quadrantal arc} \end{cases}$

    **b**  $\cos \dfrac{\theta}{2} = \begin{cases} \sqrt{\dfrac{1 + \cos \theta}{2}} & \dfrac{\theta}{2} \text{ a first or fourth quadrantal arc} \\[3ex] -\sqrt{\dfrac{1 + \cos \theta}{2}} & \dfrac{\theta}{2} \text{ a second or third quadrantal arc} \end{cases}$

**3** Use parts (a) and (b) of Prob. 2 to show that

    **a**  $\tan \dfrac{\theta}{2} = \begin{cases} \sqrt{\dfrac{1 - \cos \theta}{1 + \cos \theta}} & \dfrac{\theta}{2} \text{ a first or third quadrantal arc} \\[3ex] -\sqrt{\dfrac{1 - \cos \theta}{1 + \cos \theta}} & \dfrac{\theta}{2} \text{ a second or fourth quadrantal arc} \end{cases}$

    **a′**  $\tan \dfrac{\theta}{2} = \dfrac{1 - \cos \theta}{\sin \theta}$    for all $\theta$ for which $\tan \dfrac{\theta}{2}$ exists

    **a″**  $\tan \dfrac{\theta}{2} = \dfrac{\sin \theta}{1 + \cos \theta}$    for all $\theta$ for which $\tan \dfrac{\theta}{2}$ exists

Hint: For part (a′)

$$\sin^2 \frac{\theta}{2} = \frac{1 - \cos \theta}{2} \qquad \text{from part (a) of Prob. 2}$$

$$\sin \frac{\theta}{2} \cos \frac{\theta}{2} = \frac{\sin \theta}{2} \qquad \text{from part (b) of Prob. 1}$$

Divide.

Hint: For part (a″): Multiply (a′) by $\dfrac{1 + \cos \theta}{1 + \cos \theta}$ and simplify.

**4**  Use (I) to show that

$$\sin x \cos y = \tfrac{1}{2} \sin (x - y) + \tfrac{1}{2} \sin (x + y)$$

**5**  Use (II) to show that

**a**  $\sin x \sin y = \tfrac{1}{2} \cos (x - y) - \tfrac{1}{2} \cos (x + y)$

**b**  $\cos x \cos y = \tfrac{1}{2} \cos (x - y) + \tfrac{1}{2} \cos (x + y)$

**6**  Set $x + y = A$ and $x - y = B$ and use the identities of Probs. 4 and 5 above to derive the identities:

**a**  $\sin A + \sin B = 2 \sin \tfrac{1}{2}(A + B) \cos \tfrac{1}{2}(A - B)$

**b**  $\sin A - \sin B = 2 \cos \tfrac{1}{2}(A + B) \sin \tfrac{1}{2}(A - B)$

**c**  $\cos A + \cos B = 2 \cos \tfrac{1}{2}(A + B) \cos \tfrac{1}{2}(A - B)$

**d**  $\cos A - \cos B = -2 \sin \tfrac{1}{2}(A + B) \sin \tfrac{1}{2}(A - B)$

**7**  Show that for some values of x it is not true that

**a**  $\tfrac{1}{2} \sin 2x = \sin x$      **b**  $2 \cos x/2 = \cos x$

*In Probs. 8 and 9 compute the value of sine, cosine, and tangent of each of the following arcs, leaving your answer in radical form.*

**8**  **a**  $\tfrac{1}{12}\pi$    **b**  $\tfrac{5}{12}\pi$    **c**  $\tfrac{7}{12}\pi$    **d**  $\tfrac{11}{12}\pi$

  **e**  $\tfrac{13}{12}\pi$    **f**  $\tfrac{17}{12}\pi$    **g**  $\tfrac{19}{12}\pi$    **h**  $\tfrac{23}{12}\pi$

**9**  **a**  $\tfrac{1}{8}\pi$    **b**  $\tfrac{3}{8}\pi$    **c**  $\tfrac{5}{8}\pi$    **d**  $\tfrac{7}{8}\pi$

  **e**  $\tfrac{9}{8}\pi$    **f**  $\tfrac{11}{8}\pi$    **g**  $\tfrac{13}{8}\pi$    **h**  $\tfrac{15}{8}\pi$

**10**  Compute the value of each of the six trigonometric functions for the following arcs, leaving your answer in radical form.

  **a**  $\tfrac{1}{24}\pi$    **b**  $\tfrac{1}{16}\pi$    **c**  $\tfrac{5}{24}\pi$    **d**  $-\tfrac{1}{6}\pi$

## 9.9  IDENTITIES

Most of the exercises of Sec. 9.8 are identities, that is, equations that are true for every value of the variable or variables present (for which the functions are defined). Those of Probs. 1 to 6 inclusive are generally considered basic in a thorough and complete course in trigonometry, and often the student is asked to memorize them. These and many other similar identities arise quite naturally in pure and applied mathematics and in engineering. Although it is not our main purpose to spend much time on more identities, it is worthwhile to work through a few in order to fix more firmly in mind the fundamental relations existing among the trigonometric functions. To this end we illustrate with some examples.

◆ **Illustration 9.1**

Solve the identity

$$\frac{\sin^2 x}{1 - \cos x} = 1 + \cos x$$

provided $1 - \cos x \neq 0$.

**Solution**

Since $\sin^2 x = 1 - \cos^2 x$, we have

$$\frac{1 - \cos^2 x}{1 - \cos x} = 1 + \cos x$$

that is, factoring,

$$\frac{(1 - \cos x)(1 + \cos x)}{1 - \cos x} = 1 + \cos x$$

or, finally,

$$1 + \cos x = 1 + \cos x$$

provided $1 - \cos x \neq 0$. ◀

◆ **Illustration 9.2**

Solve the identity

$$1 + \tan^2 x = \sec^2 x \qquad x \neq \frac{(2n + 1)\pi}{2}$$

(The tangent function does not exist for an odd multiple of $\pi/2$).

**Solution**

From the pythagorean theorem $\sin^2 x + \cos^2 x = 1$. Divide by $\cos^2 x$, $x \neq (2n + 1)\pi/2$. Thus

$$\frac{\sin^2 x}{\cos^2 x} + 1 = \frac{1}{\cos^2 x}$$

or

$$\tan^2 x + 1 = \sec^2 x$$

(Similarly, $1 + \cot^2 x = \csc^2 x$, $x \neq n\pi$.) ◀

**PROBLEMS 9.9**

*Prove the following identities.*

**1** $\sin x(\cot x + \csc x) = \cos x + 1$

**2** $\dfrac{\sin x \sec x}{\tan x + \cot x} = 1 - \cos^2 x$

**3** $\tan x + \cot x = \sec x \csc x$

**4** $\dfrac{1 + \tan x}{1 - \tan x} + \dfrac{1 + \cot x}{1 - \cot x} = 0$

**5** $\dfrac{\sec x}{\csc x} = \dfrac{1 + \tan x}{1 + \cot x}$

**6** $\dfrac{\sin x + \cos x}{\sec x + \csc x} = \dfrac{\sin x}{\sec x}$

**7** $\dfrac{1 - \sin x}{\cos x} = \dfrac{\cos x}{1 + \sin x}$

**8** $\dfrac{1 + \cos x}{\sin x} + \dfrac{\sin x}{1 + \cos x} = 2 \csc x$

*In the following identities, hints in parentheses suggest the use of formulas in the problems of Sec. 9.8.*

**9** $4 \sin^2 x \cos^2 x = 1 - \cos^2 2x$    (Prob. 1)

**10** $\dfrac{1 - \cos 2x}{\sin 2x} = \tan x$    (Prob. 1)

**11** $\csc x - \cot x = \tan \frac{1}{2}x$    (Prob. 3)

**12** $\sin^2 7x + \cos 14x = \cos^2 7x$    (Prob. 1)

**13** $\cos (x + y) \cos y + \sin (x + y) \sin y = \cos x$

**14** $\sin x = \cos x \tan x = \dfrac{\tan x}{\sec x}$

**15** $\sec x + \tan x = \tan \left( \dfrac{x}{2} + \dfrac{\pi}{4} \right)$

**16** $\csc x = \cot x + \tan \dfrac{x}{2}$

**17** $\sin 4x \cos 3x + \cos 4x \sin 3x = \sin 7x$

**18** $\cos \frac{3}{5}x \cos \frac{2}{5}x - \sin \frac{2}{5}x \sin \frac{3}{5}x = \cos x$

## 9.10 TRIGONOMETRIC EQUATIONS

Equations such as

$$1 - \cos x = 0 \qquad\qquad (1)$$
$$\tan x - \cot x = 0 \qquad\qquad (2)$$
$$\csc x - 2 = 0 \qquad\qquad (3)$$

are quite evidently not identities but are conditional equalities. Equation (1) is true if and only if $x = 2n\pi$, where $n$ is an integer; equations

(2) and (3) are satisfied if and only if $x = \pi/4 + n(\pi/2)$, where $n$ is an integer. A given equation might have no solution; $\sin x = 3$ is an example. In case an equation is complicated, we may not be able to tell offhand whether it is a conditional equation or an identity.

There are practically no general rules which, if followed, will lead to the roots of a trigonometric equation. You might try to factor or to solve by quadratic formula where appropriate. Or, again, you might reduce each and every trigonometric function present to one and the same function of one and the same independent variable. In this section we exhibit some of the obvious ways of solving such an equation.

### Illustration 9.3

Solve the equation

$$2 \sin^2 x + \sin x - 1 = 0$$

for all roots.

### Solution

The left-hand member is quadratic in the quantity $\sin x$; that is, it is a polynomial of the second degree in $\sin x$. It is factorable:

$$(2 \sin x - 1)(\sin x + 1) = 0$$

Thus from the first factor we get

$$2 \sin x - 1 = 0$$
$$\sin x = \tfrac{1}{2}$$

$$x = \frac{\pi}{6} + 2n\pi \qquad \text{First quadrantal arc}$$

$$x = \tfrac{5}{6}\pi + 2n\pi \qquad \text{Second quadrantal arc}$$

The second factor yields

$$\sin x + 1 = 0$$
$$\sin x = -1$$
$$x = \tfrac{3}{2}\pi + 2n\pi$$

There are no other roots.

### Illustration 9.4

Find all values of $x$ in the interval 0 to $2\pi$ satisfying the equation

$$\cos^2 2x + 3 \sin 2x - 3 = 0$$

## Solution

This appears to offer some difficulty at first thought, because of the presence of both sine and cosine. We use the identity $\sin^2 2x + \cos^2 2x = 1$ and rewrite the equation in the form

$$1 - \sin^2 2x + 3 \sin 2x - 3 = 0$$

which factors into

$$(1 - \sin 2x)(2 - \sin 2x) = 0$$

The first factor yields

$$1 - \sin 2x = 0$$
$$\sin 2x = 1$$
$$2x = \frac{\pi}{2} + 2n\pi$$

Whence
$$x = \frac{\pi}{4} + n\pi$$

The second factor leads to the equation

$$\sin 2x = 2$$

which has no solution. ◀

▶ **Illustration 9.5**

Solve the equation $\tan^2 x - 5 \tan x - 4 = 0$.

## Solution

This is a quadratic equation in the quantity $\tan x$. Solving this by formula, we get

$$\tan x = \frac{5 \pm \sqrt{25 + 16}}{2}$$
$$= \tfrac{5}{2} \pm \tfrac{1}{2} \sqrt{41}$$
$$= 2.50000 \pm 3.20656$$
$$= 5.70656 \text{ and } -0.70656$$

Since these values do not correspond to any of the special arcs, we must resort to a table. Extensive tables exist for all the trigonometric functions. Tables V and VI, Appendix B are quite condensed, but they are adequate for our work. We find $\tan 1.39 = 5.4707$, and $\tan 1.40 = 5.7979$; hence we must interpolate. We write

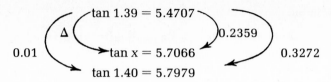

and set up the proportionality

$$\frac{\Delta}{0.01} = \frac{0.2359}{0.3272}$$

whence

$$\Delta = \frac{2359}{3272}(0.01)$$

$$\approx 0.0072$$

Therefore     $\tan(1.39 + 0.0072) = 5.7066$

and     $x = 1.3972 + 2n\pi$     First quadrantal arc

$x = (1.3972 + \pi) + 2n\pi$

$= 4.5388 + 2n\pi$     Third quadrantal arc

Now we must use $\tan x = -0.70656$; but negative values do not occur in the table. This is no handicap, however, since the sign merely tells us the arcs are second and fourth quadrantal arcs. Temporarily we write

$$\tan x' = 0.70656$$

From the table we find

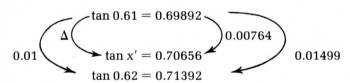

The proportionality is

$$\frac{\Delta}{0.01} = \frac{0.00764}{0.01499}$$

from which     $\Delta = \frac{764}{1499}(0.01)$

$$\approx 0.0051$$

Therefore     $\tan(0.61 + 0.0051) = 0.70656$

and     $x' = 0.6151$

But now we must use the minus sign since at present what we have is

$$\tan 0.6151 = 0.70656$$

whereas we seek x such that $\tan x = -0.70656$. This means that either

$$x = \pi - x'$$

or $$x = 2\pi - x'$$

Finally, therefore, we have

$$x = 2.5265 + 2n\pi \qquad \text{Second quadrantal arc}$$
$$x = 5.6681 + 2n\pi \qquad \text{Fourth quadrantal arc} \quad \blacklozenge$$

**PROBLEMS 9.10**

*Solve the following equations for all roots.*

|   |   |   |   |
|---|---|---|---|
| **1** | $2 \cos^2 x - \cos x = 0$ | **2** | $\sin 3x = \frac{1}{2}$ |
| **3** | $2 \sin^2 x - \sin x - 1 = 0$ | **4** | $\tan^2 x = 1$ |
| **5** | $\cos^2 x = \frac{1}{2}$ | **6** | $\csc^2 x = \frac{4}{3}$ |
| **7** | $\sec x + 1 = 2 \cos x$ | **8** | $(2 \cos x + \sqrt{3})(\sec x - 2) = 0$ |
| **9** | $\dfrac{\csc x}{\cot x} = \tan x$ | **10** | $\cos 3x = 1$ |
| **11** | $\sin^2 x - \sin x = \frac{1}{4}$ | **12** | $\cos^2 x + \cos x = \frac{1}{2}$ |

**9.11    INVERSE TRIGONOMETRIC FUNCTIONS**

When we write $y = \sin x$, we mean that given x we can find y, or
"y is the sine of x." But of course, saying that "y is the sine of x"
is the same as saying that "x is a real number whose sine is y." In
this case we regard y as given and hence determine x.

    This process should be recognized as that of forming the inverse
of the function $y = \sin x$ (Chap. 7). You will recall that the inverse
of $y = f(x)$ was obtained by first switching variables, getting $x = f(y)$,
and then solving this for $y = f^{-1}(x)$. In the present case we have to
invent a new name for the inverse function $f^{-1}$ and also to restrict the
domain so that an inverse function is defined. Let the domain of sine
be restricted to $-\pi/2 \leq x \leq \pi/2$. This function we shall indicate by
writing $y = \text{Sin } x$, using a capital S, read "y equals Cap-Sin x." We
write its inverse as $y = \text{arc Sin } x$ and read "y equals arc Cap-Sin x,"
or "y equals inverse Cap-Sin x." When y is thought of as an angle,
we may read this, "y is the angle whose Cap-Sin is x." Sometimes the
notation $y = \text{Sin}^{-1} x$ is used.

    The two functions whose values are given by $\text{Sin } x$ and $\text{arc Sin } x$
are different functions; either is said to be the inverse of the other.
We sketch in Fig. 9.15 the graph of $y = \text{Sin } x$ and that of $y = \text{arc Sin } x$.

The domain and range of $y = \mathrm{Sin}\, x$ are $-\pi/2 \le x \le \pi/2$, $-1 \le y \le 1$. The domain and range of $y = \mathrm{arc\ Sin}\, x$ are $-1 \le x \le 1$, $-\pi/2 \le y \le \pi/2$. Similarly we define other restricted trigonometric functions.

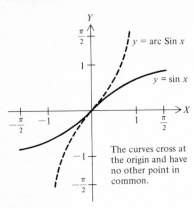

**Figure 9.15**

$y = \mathrm{Cos}\, x$, $0 \le x \le \pi$, $-1 \le y \le 1$, as cos x restricted to this domain
$y = \mathrm{arc\ Cos}\, x$, $-1 \le x \le 1$, $0 \le y \le \pi$, as the inverse of $y = \mathrm{Cos}\, x$
$y = \mathrm{Tan}\, x$, $-\pi/2 < x < \pi/2$, $-\infty < y < \infty$, as tan x restricted to this domain
$y = \mathrm{arc\ Tan}\, x$, $-\infty < x < \infty$, $-\pi/2 < y < \pi/2$, as the inverse of $y = \mathrm{Tan}\, x$

## PROBLEMS 9.11

1  Find the value of the following.

   **a**  $\mathrm{Sin}^{-1} \tfrac{1}{2}\sqrt{2}$    **b**  $\mathrm{arc\ Cos}\left(\dfrac{-\sqrt{3}}{2}\right)$

   **c**  $\mathrm{Tan}^{-1} 0$    **d**  arc Tan (sin 270°)

2  Find the value of the following.

   **a**  $y = \mathrm{arc\ Sin}\, \tfrac{1}{2}$    **b**  $y = \mathrm{arc\ Sin}\, 0$

   **c**  $y = \mathrm{Cos}^{-1}(-1)$    **d**  $y = \mathrm{Tan}^{-1} 5.1446$

3  Verify the following.

   **a**  $\mathrm{Sin}^{-1} \tfrac{1}{2} + \mathrm{Sin}^{-1} \tfrac{1}{2}\sqrt{3} = -\mathrm{Sin}^{-1}(-1)$

   **b**  $\mathrm{Cos}^{-1} x = \mathrm{Tan}^{-1}(\sqrt{1 - x^2}/x)$, $0 < x \le 1$

4  Find the value of the following.

   **a**  Cos (arc Sin $\tfrac{3}{5}$)    **b**  csc (arc Sin $\tfrac{3}{5}$)    **c**  sec (Tan$^{-1}$ 2)

    **d** $\mathrm{Tan}^{-1}(\mathrm{Tan}\ 5)$      **e** $\mathrm{Sin}\ (\mathrm{arc}\ \mathrm{Cos}\ \frac{4}{5})$      **f** $\cos\ (\mathrm{arc}\ \mathrm{Tan}\ \frac{2}{3})$

**5** Verify the following.

$$2\ \mathrm{Tan}^{-1}\tfrac{1}{3} + \mathrm{Tan}^{-1}\tfrac{1}{7} = \frac{\pi}{4}$$

**6** Sketch the graph of the following.

    **a** $y = \mathrm{arc}\ \mathrm{Cos}\ x$      **b** $y = \mathrm{arc}\ \mathrm{Tan}\ x$

    **c** $y = \mathrm{Cos}\ x$      **d** $y = \mathrm{Tan}\ x$

## 9.12 TRIGONOMETRY OF ANGLES

We have defined the trigonometric functions in terms of a real independent variable $S$; this real number has been taken as the length of an arc of the unit circle. Now suppose we consider a circle of radius $r\ (\neq 0)$ and center $O$ on which there is given an arc $PQ$ which is directed from $P$ to $Q$. There is associated with each such arc a certain directed angle. Draw the infinite rays (half-lines) $p$ and $q$ through $OP$ and $OQ$, respectively (Fig. 9.16). Now rotate $p$ about $O$ into $q$ so that $P$ traverses the arc $PQ$. The rotation may be counterclockwise, as in Fig. 9.16a, or clockwise, as in Fig. 9.16b. Moreover, it may include one or more complete revolutions, as in Fig. 9.16c. In order to describe this rotation, we must know not only the positions of $p$ and $q$, but also the arc $PQ$ which tells us how $p$ is rotated into $q$. We can now define a directed angle.

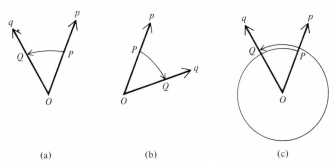

    (a)                (b)              (c)

**Figure 9.16**

## DEFINITIONS 9.5

A *directed angle* is the rotation of a ray $p$ into a ray $q$ in such a way that $P$ traverses a given arc $PQ$. We call $p$ the *initial side* and $q$ the *terminal side* of this directed angle. ◆

If we choose another point $P'$ on $p$ and a point $Q'$ on $q$ (Fig. 9.17) such that $OP' = OQ' = r'$, the above rotation will send $P'$ into $Q'$ along an arc of a circle of radius $r'$. This will therefore generate a directed angle associated with the arc $P'Q'$. Directed angles which are associated with arcs $PQ$ and $P'Q'$, which are related in this fashion, will be said to be identical. We will use letters $\alpha$, $\beta$, $\gamma$, . . . , and $A$, $B$, $C$, . . . to represent directed angles.

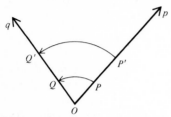

**Figure 9.17**

It is now possible to speak of the measure of a directed angle. For any such measure we must have a suitable *unit* for our measurements. There are a number of these in common use, the simplest of which for theoretical purposes is the *radian*. In order to define the radian measure of a directed angle, choose the arc $PQ$ of radius 1 with which it is associated.

### DEFINITION 9.6

*The measure in radians* of a directed angle is the length (together with its algebraic sign) of the arc of a unit circle, $PQ$ with which the angle is associated.  ◀

We recall that arcs running counterclockwise are positive and that arcs running clockwise are negative. Suppose that we have two arcs $PQ$ and $P'Q'$ which are on circles of radius 1 and $r$, respectively, and which are associated with the same directed angle. Then from plane geometry we know that

$$\frac{\text{Length of arc } P'Q'}{\text{Length of arc } PQ} = \frac{r}{1}$$

Since the length of arc $PQ$ is equal to the measure of $\theta$ in radians, it follows that

$$\text{Length of arc } P'Q' = r\theta$$

This formula is usually written

$$S = r\theta$$

where $S$ stands for the length of arc $P'Q'$.

The most customary measure of angle in practical situations is the familiar *degree,* "°." By the definition of a degree, there are 360° in a complete counterclockwise revolution; that is, a revolution is divided into 360 equal parts, each of which has the measure of 1°. Each degree is further divided into 60 minutes (60′) and each minute into 60 seconds (60″). Thus $2\pi$ radians are equivalent to 360°. Hence

$$1 \text{ radian} = \left(\frac{360}{2\pi}\right)^{\circ} \qquad 1^{\circ} = \left(\frac{2\pi}{360}\right) \text{ radians}$$

$$= \left(\frac{180}{\pi}\right)^{\circ} \qquad = \left(\frac{\pi}{180}\right) \text{ radians}$$

These relations permit us to convert degrees into radians, and vice versa. For example,

**1**  $50^{\circ} = 50\left(\frac{\pi}{180}\right) \text{ radians} = \left(\frac{5\pi}{18}\right) \text{ radians}$

$= 0.873 \text{ radians}$

**2**  $2 \text{ radians} = 2\left(\frac{180}{\pi}\right)^{\circ} = 114.5913^{\circ} = 114^{\circ}35'29''$

Convenient tables for making this conversion are available.

A directed angle will be said to be in *standard position* when the initial side $p$ is along the positive $X$ axis. Clearly any directed angle can be rotated into standard position. To find the trigonometric functions of a directed angle, we first put it into standard position and then use the correspondence between directed angles and arcs.

## DEFINITION 9.7

Let $\alpha$ be a directed angle in standard position associated with an arc S of a unit circle. Then $\sin \alpha$, $\cos \alpha$, etc., are defined to be the real numbers $\sin$ S, $\cos$ S, etc., defined in Sec. 9.3.  ◀

This definition permits us to define $\sin \alpha$ and $\cos \alpha$ directly as follows. Place the directed angle $\alpha$ in standard position with its vertex at the origin $O$. Draw a unit circle with center $O$, and let its intersection with the terminal side $q$ of $\alpha$ be $Q$. Then the x and y coordinates of $Q$ are $\cos \alpha$ and $\sin \alpha$, respectively (Fig. 9.18).

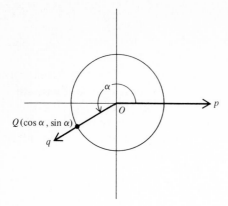

**Figure 9.18**

Since every directed angle is completely determined by its measure, we may write sin 2 radians or sin 114°35′29″ in place of sin α when the measure of α is 2 radians = 114°35′29″.

Thus we may speak of

$$\sin 90° = \sin \frac{\pi}{2} \text{ radians} = \sin \frac{\pi}{2} \qquad \text{by definition}$$

$$\cos 180° = \cos \pi \text{ radians} = \cos \pi \qquad \text{by definition}$$

$$\tan 16° = \tan \left(16 \times \frac{\pi}{180}\right) \text{ radians} = \tan \frac{4\pi}{45} \qquad \text{by definition}$$

Tables of these functions for angles expressed in degrees are available in many places. These tables are usually labeled "natural trigonometric functions" to distinguish them from tables of the logarithms of these functions.

Where no symbol such as °, ′, or ″ is used, the angle is to be thought of as expressed in radian measure. Thus sin 7 means "the sine of an angle of 7 radians" or "the sine of an arc (of a unit circle) whose length is 7 units," the two being equivalent. But sin 7° means "the sine of an angle of 7 degrees."

## PROBLEMS 9.12

*In Probs. 1 to 4, indicate why:*

**1** The measure of a quarter revolution in the counterclockwise direction is π/2 radians.

**2**   The measure of a half revolution in the counterclockwise direction is $\pi$ radians.

**3**   The measure of a full revolution in the counterclockwise direction is $2\pi$ radians.

**4**   The measure of a full revolution in the clockwise direction is $-2\pi$ radians.

**5**   Convert the following radians to (the nearest number of) degrees:

     **a**   30     **b**   45     **c**   3     **d**   1

     **e**   $\frac{1}{2}$     **f**   $-\frac{1}{3}$     **g**   $-0.55$     **h**   0.26

**6**   Convert the following degrees to radians:

     **a**   $30°$     **b**   $45°$     **c**   $60°$     **d**   $100°$

     **e**   $225°$     **f**   $-140°$     **g**   $-80°$     **h**   $600°$

*In Probs. 7 to 10, use the appropriate table in Appendix B to find:*

**7**   The sine of each of the following angles:

     **a**   $49°$     **b**   $127°$     **c**   $233°$

     **d**   $306°$     **e**   0.2793     **f**   1.0123

**8**   The cosine of each of the following angles:

     **a**   $6°$     **b**   $160°$     **c**   $247°$

     **d**   $285°$     **e**   0.5585     **f**   1.3963

**9**   The tangent of each of the following angles:

     **a**   $91°$     **b**   $170°$     **c**   $220°$

     **d**   $330°$     **e**   0.7679     **f**   1.2217

**10**   The indicated number:

     **a**   $\sec 19°$     **b**   $\csc 72°$     **c**   $\cot 160°$

     **d**   $\sec 0$     **e**   $\cot 0.2793$     **f**   $\csc 1.1694$

*In Probs. 11 to 13, find the angle to the nearest degree satisfying:*

**11**   **a**   $\sin x = 0.62580$       (first quadrantal angle)
        **b**   $\sin x = 0.49872$       (second quadrantal angle)
        **c**   $\sin x = -0.87023$     (third quadrantal angle)
        **d**   $\sin x = -0.21319$     (fourth quadrantal angle)
**12**   **a**   $\cos x = 0.26135$       (first quadrantal angle)
        **b**   $\cos x = -0.74963$     (second quadrantal angle)
        **c**   $\cos x = -0.30903$     (third quadrantal angle)
        **d**   $\cos x = 0.55553$       (fourth quadrantal angle)
**13**   **a**   $\tan x = 6.2358$        (first quadrantal angle)
        **b**   $\tan x = -0.12343$     (second quadrantal angle)
        **c**   $\tan x = 1.4729$        (third quadrantal angle)
        **d**   $\tan x = -0.3184$      (fourth quadrantal angle)

## 9.13    RIGHT TRIANGLES

The simplest applications of trigonometry involve the solution of right triangles. We are given any two of the following parts (including at least one side): the angles $A$ and $B$, and the sides $a$, $b$, and $c$. Then we are asked to find the remaining three parts. (See Fig. 9.19.)

This problem can easily be solved using the trigonometric functions. Along $AB$ lay off a segment $AD$ of length 1. Draw $DE$ perpendicular to $AC$, and with $A$ as center draw the circular arc $FD$. Angle $A$ is now a directed angle whose measure is positive. From the definitions of the functions,

$$\sin A = DE$$
$$\cos A = AE$$
$$\tan A = \frac{DE}{AE}$$

Since triangles $ADE$ and $ABC$ are similar,

$$\frac{DE}{1} = \frac{BC}{AB} = \frac{a}{c}$$
$$\frac{AE}{1} = \frac{AC}{AB} = \frac{b}{c}$$
$$\frac{DE}{AE} = \frac{BC}{AC} = \frac{a}{b}$$

Therefore
$$\sin A = \frac{a}{c}$$
$$\cos A = \frac{b}{c}$$
$$\tan A = \frac{a}{b}$$

For easy reference these may be stated: In a right triangle,

$$\text{The sine of an acute angle} = \frac{\text{opposite side}}{\text{hypotenuse}}$$

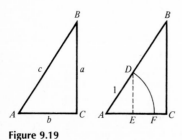

**Figure 9.19**

$$\text{The cosine of an acute angle} = \frac{\text{adjacent side}}{\text{hypotenuse}}$$

$$\text{The tangent of an acute angle} = \frac{\text{opposite side}}{\text{adjacent side}}$$

These relations were originally developed for acute angles in standard position (Fig. 9.18). Since any acute angle can be rotated into standard position, the relations hold for all acute angles.

As an illustration of the application of these formulas, consider the following problem.

◆ **Illustration 9.6**

A gable roof has rafters 10 feet long. If the eaves are 17 feet apart, how much headroom is there in the center of the attic and how steep is the roof?

**Solution**

Directly from Fig. 9.20 we see that

$$\text{Headroom} = x = \sqrt{100 - 72.25}$$
$$= \sqrt{27.75}$$
$$= 5.26 \text{ feet} \qquad \text{from a table of square roots}$$

The steepness of the roof is measured by $\theta$, where

$$\cos \theta = \frac{8.5}{10} = 0.85$$

and, from a table of natural cosines (degree measure),

$$\theta = 31°47.3' \text{ to the nearest } \tfrac{1}{10} \text{ minute} \quad ◀$$

An example of the second type is given in the following illustration.

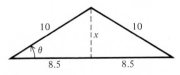

**Figure 9.20**

◆ **Illustration 9.7**

A radio tower stands on top of a building 200 feet high. At a point on the ground 500 feet from the base of the building the tower subtends an angle of 10°. How high is the tower?

**Solution**

From Fig. 9.21 we have

$$\tan \theta = \tfrac{200}{500} = 0.40000$$
$$\theta = 21°48.1' \text{ to the nearest } \tfrac{1}{10} \text{ minute}$$

Now
$$\tan 31°48.1' = \frac{200 + x}{500}$$
$$0.62007 = \frac{200 + x}{500}$$
$$200 + x = 310.04$$
$$x = 110.04 \text{ feet} \blacktriangleleft$$

There is no end to the mathematical subject called trigonometry but we have covered the basic material needed for further work in mathematics, especially in the calculus.

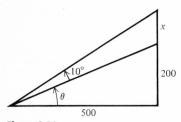

**Figure 9.21**

## PROBLEMS 9.13

**1**   From the top of a lighthouse 70 feet high the angle of depression of a boat is found to be 35°; how far away from the base of the lighthouse is the boat?

**2**   What angle does the diagonal of the face of a cube make with a diagonal of the cube (drawn from the same vertex)?

**3   a**   What is the perimeter of a regular pentagon inscribed in a circle of radius 5 inches?      **b**   Circumscribed?

**4**   A wheel 3 feet in diameter is driven, by means of a (noncrossed) belt, by a wheel 1 foot in diameter. If the wheel centers are 8 feet apart, how long is the belt?

**5**   A 10-foot ladder, with its foot anchored in an alleyway, will reach 9 feet up a building on one side of the alley and 6 feet up a building on the other. How wide is the alley?

**6**   Discover a way of measuring the width of a stream by making all the measurements on one bank.

# 10

## APPLICATIONS
## OF TRIGONOMETRY
## TO GRAPHING

**10.1    INTRODUCTION**

In Chap. 9 you were introduced to some new functions that have interesting properties and interrelationships. These trigonometric or circular functions were defined by means of the coordinates of points on a circle and, therefore, are particularly compatible with circular or recurring phenomena. In this chapter we will take advantage of trigonometric functions to develop two different approaches to graphing.

One of the approaches enables us to represent locations in terms of *polar coordinates* instead of the rectangular coordinates that we have used up to now. Polar coordinates prove to be particularly compatible with the graphs of certain geometric figures.

In the other approach we will use parametric equations to express relationships which can convey more information than can single functions. Parametric equations can be graphed in both polar and rectangular coordinates.

## 10.2   POLAR COORDINATES

Figure 10.1a includes all of the essential ingredients in describing the location of a point in polar coordinates. One starts with a point $O$ (called the origin) and a half-infinite line (called the axis) extending from $O$. On the axis one establishes a unit of length. Then one locates any point $P$ by means of the ordered pair $(r, \theta)$, where $r$ is a directed distance and $\theta$ is the measure of an angle. The various ways of choosing $r$ and $\theta$ will be illustrated in the examples which follow.

### ◗ Illustration 10.1

See Fig. 10.1b. One choice of $(r, \theta)$ for $P$ in Fig. 10.1a is $(\frac{3}{2}, \pi/4)$, since the positive distance from $O$ to $P$ is $\frac{3}{2}$, and since the counterclockwise angle between the axis and $\overline{OP}$ is $\pi/4$ (Fig. 10.1c).

One of the useful but confusing features of polar coordinates is that they are not unique. That is, there are other pairs $(r, \theta)$ which can describe the location of this point $P$ (Fig. 10.1d).

In this illustration you can see that, if you rotate through the angle $5\pi/4$, then the point $P$ is behind you at a distance of $\frac{3}{2}$, that is, it has a directed distance of $-\frac{3}{2}$. Hence $(-\frac{3}{2}, 5\pi/4)$ also describes the location of $P$. So do $(-\frac{3}{2}, -3\pi/4)$, $(\frac{3}{2}, -7\pi/4)$, $(\frac{3}{2}, \pi/4 + 2\pi)$, $(-\frac{3}{2}, 5\pi/4 + 4\pi)$, etc., as in Figs. 10.1e–h. It should be clear from these illustrations that the angle $\theta$ can be chosen in an infinite number of ways, and for each of these ways one of two possible choices of $r$ is appropriate. ◗

### ◗ Illustration 10.2

Figure 10.2a–f shows six ways of describing the location of a point.

In Fig. 10.3 we locate each of the following points: $(-2, 0)$, $(2, 2\pi)$, $(\pi, \pi)$, $(1, -\pi/6)$, $(1, 1)$. It may be worth mentioning that the point $(1, 1)$ was located by estimating $\theta = 1$ radian. ◗

### PROBLEMS 10.2

1   Give eight different ordered pairs $(r, \theta)$ which describe the location of the point $P$ relative to the origin and axis pictured in Fig. 10.4.

2   Give eight different ordered pairs $(r, \theta)$ which describe the location of the point $P$ relative to the origin and axis pictured in Fig. 10.5.

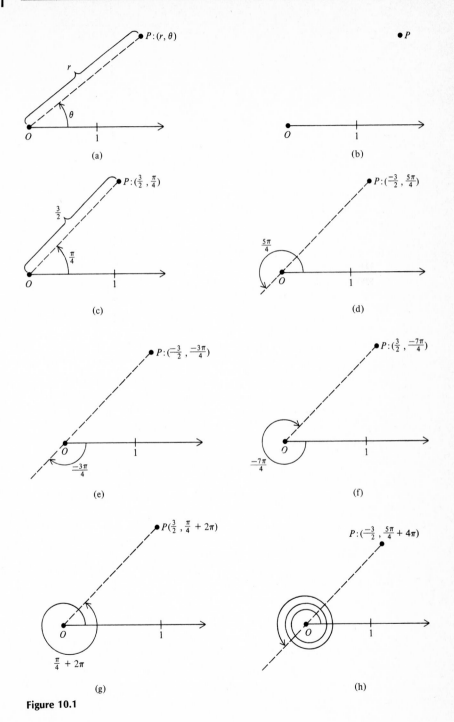

**Figure 10.1**

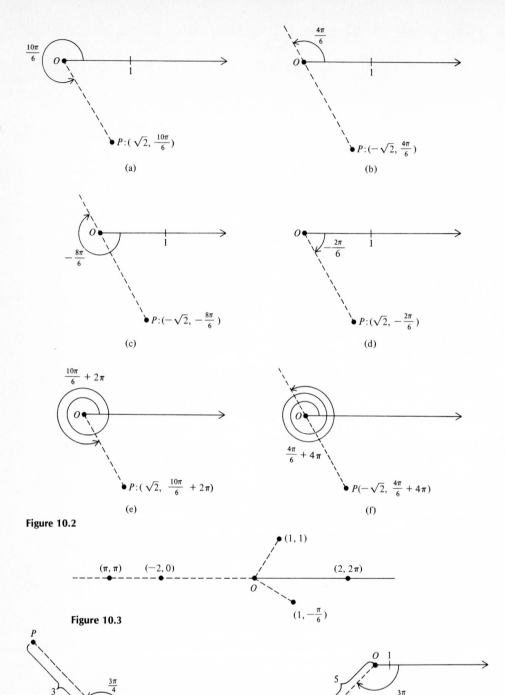

$\frac{10\pi}{6}$

$O$

1

$P:(\sqrt{2}, \frac{10\pi}{6})$

(a)

$\frac{4\pi}{6}$

$O$

1

$P:(-\sqrt{2}, \frac{4\pi}{6})$

(b)

$O$

1

$-\frac{8\pi}{6}$

$P:(-\sqrt{2}, -\frac{8\pi}{6})$

(c)

$O$

$-\frac{2\pi}{6}$

1

$P:(\sqrt{2}, -\frac{2\pi}{6})$

(d)

$\frac{10\pi}{6} + 2\pi$

$O$

$P:(\sqrt{2}, \frac{10\pi}{6} + 2\pi)$

(e)

$O$

$\frac{4\pi}{6} + 4\pi$

$P(-\sqrt{2}, \frac{4\pi}{6} + 4\pi)$

(f)

**Figure 10.2**

$(1, 1)$

$(\pi, \pi)$     $(-2, 0)$          $(2, 2\pi)$

$O$

$(1, -\frac{\pi}{6})$

**Figure 10.3**

$P$

3

$\frac{3\pi}{4}$

$O$   1

**Figure 10.4**

$O$   1

5

$-\frac{3\pi}{4}$

$P$

**Figure 10.5**

**3**  Give four different ordered pairs of polar coordinates for each of the points described below in polar coordinates.

    **a** $(1, 2)$     **b** $(-3/2, 2\pi/3)$     **c** $(3, -\pi)$     **d** $(0, 0)$

**4**  Give four different ordered pairs of polar coordinates for each of the points described below in polar coordinates.

    **a** $(2, 3)$     **b** $(-5, \pi/4)$     **c** $(1, -3\pi/2)$     **d** $(\sqrt{2}, 0)$

**5**  How many different $(r, \theta)$ with $0 \le \theta \le 2\pi$, $r \ne 0$ can be used to describe a single point $P$? Describe why your answer is true.

**6**  Locate the points which have the following polar coordinates.

    **a** $(1, 0)$     **b** $(\sqrt{3}, -\pi)$     **c** $(-\sqrt{3}, \pi)$

    **d** $(4, 3)$     **e** $(4, 9\pi/4)$     **f** $(4, \pi/4)$

**7**  Locate the points which have the following polar coordinates with respect to a single origin, axis, and unit.

    **a** $(0, 1)$     **b** $(\pi, -\pi/2)$     **c** $(3, 5\pi/6)$

    **d** $(3, 4)$     **e** $(1, 11\pi/4)$     **f** $(-7/2, -3\pi/2)$

**8**  Choose the correct one of the following statements and give examples to illustrate the chosen statement.

    **a**  Each pair $(r, \theta)$ describes many different points.

    **b**  Each point is represented by many different pairs $(r, \theta)$.

**9**  Recall that for rectangular cartesian coordinates graph paper is made by drawing in the lines $x =$ constant and $y =$ constant for certain constant values. The result is a coordinate grid which looks like Fig. 10.6. What would the grid for polar coordinates look like if you constructed it by drawing the curves $r =$ constant and $\theta =$ constant?

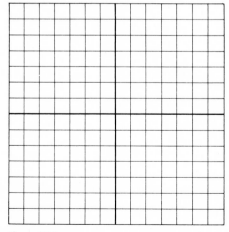

**Figure 10.6**

**10**   Describe a set of rules for locating points using polar coordinates so that each point has a unique pair of coordinates.

## 10.3   GRAPHING POLAR EQUATIONS

Just as you graphed equations such as $y = x$, $x^2 + y^2 = 4$, $y = \sin x$, and $y = e^x$ with respect to rectangular coordinates, you can graph equations such as $r = \theta$, $r^2 + \theta^2 = 4$, $r = \sin \theta$, and $r = e^\theta$ with respect to polar coordinates. You will find that the graphs look quite different, and you will find that certain equations are much more compatible with one coordinate system than they are with the other.

◆   **Illustration 10.3**

Graph $r = \theta$, $\theta \geq 0$.

| $\theta$ | 0 | $\dfrac{\pi}{6}$ | $\dfrac{\pi}{4}$ | $\dfrac{\pi}{2}$ | $\pi$ | $\dfrac{3\pi}{2}$ | $2\pi$ |
|---|---|---|---|---|---|---|---|
| $r$ | 0 | $\dfrac{\pi}{6}$ | $\dfrac{\pi}{4}$ | $\dfrac{\pi}{2}$ | $\pi$ | $\dfrac{3\pi}{2}$ | $2\pi$ |

You can see from Fig. 10.7 that the graph goes on and on, spiraling outward. As with rectangular coordinates, you should plot as many points as you need in order to be confident of the shape of the graph. You will probably want to choose values for $\theta$ which are easy for you to represent. [What does $r = \theta$ ($\theta \leq 0$) look like?]   ◆

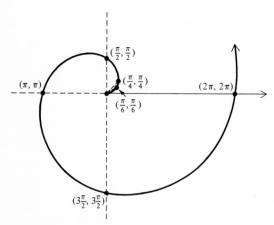

**Figure 10.7**

▶ **Illustration 10.4**

$$r = \sin \theta$$

As $\theta$ increases from 0 to $\pi/2$, $r(=\sin\theta)$ takes on values between 0 and 1 as indicated by the following table.

| $\theta$ | 0 | $\dfrac{\pi}{6}$ | $\dfrac{\pi}{4}$ | $\dfrac{\pi}{3}$ | $\dfrac{\pi}{2}$ |
|---|---|---|---|---|---|
| $r$ | 0 | $\dfrac{1}{2}$ | $\dfrac{\sqrt{2}}{2}$ | $\dfrac{\sqrt{3}}{2}$ | 1 |

The portion of the graph which corresponds to values of $\theta$ between 0 and $\pi/2$ appears in Fig. 10.8a.

As $\theta$ increases from $\pi/2$ to $\pi$, $r$ decreases from 1 back to 0. This fact is reflected in the following table and in the graph in Fig. 10.8b.

| $\theta$ | $\dfrac{\pi}{2}$ | $\dfrac{2\pi}{3}$ | $\dfrac{3\pi}{4}$ | $\dfrac{5\pi}{6}$ | $\pi$ |
|---|---|---|---|---|---|
| $r$ | 1 | $\dfrac{\sqrt{3}}{2}$ | $\dfrac{\sqrt{2}}{2}$ | $\dfrac{1}{2}$ | 0 |

A composite drawing of the graph of $r = \sin\theta$ for $0 \le \theta \le \pi$ is given in Fig. 10.8c. The fact that the graph of $r = \sin\theta$ is symmetric about

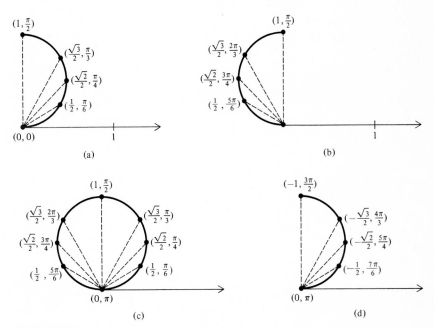

Figure 10.8

$\theta = \pi/2$ could have been anticipated from the identity $\sin(\theta - \pi/2) = \sin(\theta + \pi/2)$.

In the next section we will show that this graph of $r = \sin\theta$ for $0 \le \theta \le \pi$ is a circle with center $(\frac{1}{2}, \pi/2)$ and radius $\frac{1}{2}$.

What happens for $\pi \le \theta \le 3\pi/2$?

| $\theta$ | $\pi$ | $\dfrac{7\pi}{6}$ | $\dfrac{5\pi}{4}$ | $\dfrac{4\pi}{3}$ | $\dfrac{3\pi}{2}$ |
|---|---|---|---|---|---|
| $r$ | $0$ | $-\dfrac{1}{2}$ | $\dfrac{-\sqrt{2}}{2}$ | $\dfrac{-\sqrt{3}}{2}$ | $-1$ |

This table and Fig. 10.8d indicate that as $\theta$ increases from $\pi$ to $3\pi/2$, $r$ takes on negative values from 0 to $-1$, and so the first half of the circle is retraced. You should determine for yourself that as $\theta$ increases from $3\pi/2$ to $2\pi$, the second half of the circle is retraced. These negative values for $\sin\theta$ and the fact that the curve was retraced could have been anticipated from the identity $\sin(\theta + \pi) = -\sin\theta$. Of course, for $\theta > 2\pi$, the identity $\sin(\theta + 2\pi) = \sin\theta$ assures us that the circle will continue to be retraced.

In summary, the graph of $r = \sin\theta$ is the circle pictured in Fig. 10.8c. The circle is traced completely as $\theta$ takes on all the values in any interval of length $\pi$. ◀

▶ **Illustration 10.5**

Graph $r = \sin 2\theta$.

You will be surprised at the difference the 2 makes. For $\theta$ between 0 and $\pi$, we make a table showing $\theta$, $2\theta$, and $r$, and then we plot the corresponding points (Fig. 10.9a).

| $\theta$ | $2\theta$ | $r$ | $\theta$ | $2\theta$ | $r$ |
|---|---|---|---|---|---|
| $0$ | $0$ | $0$ | $\dfrac{\pi}{2}$ | $\pi$ | $0$ |
| $\dfrac{\pi}{12}$ | $\dfrac{\pi}{6}$ | $\dfrac{1}{2}$ | $\dfrac{5\pi}{8}$ | $\dfrac{5\pi}{4}$ | $\dfrac{-\sqrt{2}}{2}$ |
| $\dfrac{\pi}{8}$ | $\dfrac{\pi}{4}$ | $\dfrac{\sqrt{2}}{2}$ | $\dfrac{3\pi}{4}$ | $\dfrac{3\pi}{2}$ | $-1$ |
| $\dfrac{\pi}{4}$ | $\dfrac{\pi}{2}$ | $1$ | $\dfrac{7\pi}{8}$ | $\dfrac{7\pi}{4}$ | $\dfrac{-\sqrt{2}}{2}$ |
| $\dfrac{\pi}{3}$ | $\dfrac{2\pi}{3}$ | $\dfrac{\sqrt{3}}{2}$ | $\pi$ | $2\pi$ | $0$ |
| $\dfrac{3\pi}{8}$ | $\dfrac{3\pi}{4}$ | $\dfrac{\sqrt{2}}{2}$ | | | |

Now you need to start filling in values of $\theta$ between $\pi$ and $2\pi$. This could get a little tedius unless we remember how the sine function be-haves. For $\pi \le \theta < 2\pi$, $2\pi \le 2\theta < 4\pi$. Moreover we know that $\sin(\theta + 2\pi) = \sin \theta$. So we know that for $\pi \le \theta < 2\pi$, $\sin 2\theta$ repeats the values it took on for $0 \le \theta < \pi$. So we can sketch Fig. 10.9b. You should check a few of these points in order to be sure that you under-stand.

Putting these two graphs together we get Fig. 10.9c. There are two things to note here. One is that one must be particularly careful about the sign of $r$ when locating point $(r, \theta)$. The second is that con-siderable labor can be saved by taking advantage of appropriate trigo-nometric identities. ◀

It is time for you to do some graphing. The main thing for you to do here is to watch the signs and to take advantage of any work-saving relationships that arise from knowledge you already have.

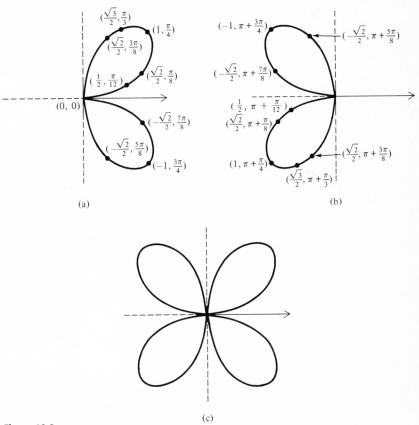

Figure 10.9

**PROBLEMS 10.3**

Graph each of the following equations.

1   $r = \dfrac{1}{\theta}$       2   $r = 3$       3   $\theta = \dfrac{\pi}{4}$

4   $r^2 = \theta$       5   $r = \cos \theta$       6   $r = \cos 2\theta$

7   $r = \cos \dfrac{\theta}{2}$       8   $r = \sin 3\theta$       9   $r = \sin \dfrac{\theta}{2}$

10   $r = 1 + 2 \sin \theta$       11   $r = 1 + 2 \cos \theta$       12   $r = 2 + \sin \theta$

13   $r = 2 + \cos \theta$       14   $r = \dfrac{1}{1 - \sin \theta}$       15   $r = \dfrac{1}{1 + \sin \theta}$

**16**   Sketch and find the points of intersection $r = \cos \theta$ and $r = 1 - \cos \theta$.

**17**   Sketch and find the points of intersection of $r = \sin \theta$ and $r = \sqrt{3} \cos \theta$.

**18**   Investigate any possible (geometric) points of intersection of the curves

$$r = \frac{1}{1 + \cos \theta}$$

$$r = \frac{-1}{1 - \cos \theta}$$

(Hint: You may want to sketch the two curves.)

**10.4**   **POLAR COORDINATES VERSUS RECTANGULAR COORDINATES**

A point $P$ can be located either by means of rectangular coordinates $(x, y)$ or by polar coordinates $\langle r, \theta \rangle$* with respect to the same origin and the positive $X$ axis as axis. In this section we will establish a relationship to help in graphing polar equations (Fig. 10.10).

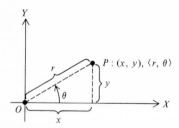

**Figure 10.10**

* Throughout the rest of this chapter the brackets $\langle$ , $\rangle$ will be used to designate polar coordinates in order to distinguish them from rectangular coordinates which will be represented in the usual way, ( , ).

From Chap. 9 we know that

$$\sin \theta = \frac{y}{r}$$

$$\cos \theta = \frac{x}{r}$$

From which we get easily

$$x = r \cos \theta$$
$$y = r \sin \theta$$

So that if you are given $\langle r, \theta \rangle$, you can find $(x, y) = (r \cos \theta, r \sin \theta)$.

◆ **Illustration 10.6**

Find the rectangular coordinates of the points with polar coordinates $\langle \frac{1}{2}, \pi/4 \rangle$, $\langle -2, -\pi/8 \rangle$, and $\langle \sqrt{3}, -\pi \rangle$.

For $\langle \frac{1}{2}, \pi/4 \rangle$,

$$x = \frac{1}{2} \cos \frac{\pi}{4} = \frac{1}{2} \cdot \frac{\sqrt{2}}{2} = \frac{\sqrt{2}}{4}$$

$$y = \frac{1}{2} \sin \frac{\pi}{4} = \frac{1}{2} \cdot \frac{\sqrt{2}}{2} = \frac{\sqrt{2}}{4}$$

So the rectangular coordinates are $(\sqrt{2}/4, \sqrt{2}/4)$.

For $\langle -2, -\pi/6 \rangle$,

$$x = -2 \cos \left(-\frac{\pi}{6}\right) = -2\left(\frac{\sqrt{3}}{2}\right) = -\sqrt{3}$$

$$y = -2 \sin \left(-\frac{\pi}{6}\right) = -2\left(-\frac{1}{2}\right) = 1$$

So the rectangular coordinates are $(-\sqrt{3}, 1)$.

For $\langle \sqrt{3}, -\pi \rangle$,

$$x = \sqrt{3} \cos (-\pi) = \sqrt{3}(-1) = -\sqrt{3}$$
$$y = \sqrt{3} \sin (-\pi) = \sqrt{3}(0) = 0$$

So the rectangular coordinates are $(-\sqrt{3}, 0)$. ◆

◆ **Illustration 10.7**

What about the reverse process? That is, what if you are given $(x, y)$ and want to find $\langle r, \theta \rangle$?

Since

$$x = r \cos \theta$$
$$y = r \sin \theta$$

we get

$$y/x = \tan \theta$$

So for $-\pi/2 < \theta < \pi/2$,

$$\theta = \text{Tan}^{-1} y/x$$

Since $\tan(\theta + n\pi) = \tan\theta$, we can conclude that for all $\theta$,

$$\theta = \text{Tan}^{-1}\, y/x + n\pi$$

We also have

$$x^2 + y^2 = r^2\cos^2\theta + r^2\sin^2\theta$$
$$= r^2(\cos^2\theta + \sin^2\theta)$$
$$= r^2(1)$$
$$= r^2$$

So

$$r = \pm\sqrt{x^2 + y^2}$$

You know that there are many different $\langle r, \theta\rangle$'s for a single $(x, y)$. So you should not be surprised to learn that you will have to choose the particular value of $\text{Tan}^{-1}\, y/x + n\pi$ and the particular sign of $\pm\sqrt{x^2 + y^2}$. ◀

▶ **Illustration 10.8**

Find polar coordinates of the points with rectangular coordinates $(1, 1)$, $(3, -\sqrt{3})$, $(7, 0)$, $(0, 7)$.

For $(1, 1)$,

$$\theta = \text{Tan}^{-1}\, 1/1 + n\pi = \text{Tan}^{-1}\, 1 + n\pi = \pi/4 + n\pi \qquad \text{for any integer } n$$
$$r = \pm\sqrt{1^2 + 1^2} = \pm\sqrt{2}$$

Then, noting the location of $(1, 1)$ in the first quadrant, we can choose any of $\langle\sqrt{2}, \pi/4 + 2n\pi\rangle$ and $\langle-\sqrt{2}, \pi/4 + (2n + 1)\pi\rangle$ for $\langle r, \theta\rangle$.

For $(3, -\sqrt{3})$,

$$\theta = \text{Tan}^{-1}\,(-\sqrt{3}/3) + n\pi = -\pi/6 + n\pi$$
$$r = \pm\sqrt{(1 - \sqrt{3})^2 + (3)^2} = \pm\sqrt{12}$$
$$= \pm 2\sqrt{3}$$

For this fourth-quadrant point we must choose from among $\langle 2\sqrt{3}, -\pi/6 + 2n\pi\rangle$ and $\langle-2\sqrt{3}, -\pi/6 + (2n + 1)\pi\rangle$ for $\langle r, \theta\rangle$.

For $(7, 0)$ we have

$$\theta = \text{Tan}^{-1}\,\tfrac{0}{7} + n\pi = \text{Tan}^{-1}\, 0 + n\pi = 0 + n\pi$$
$$r = \pm\sqrt{7^2 + 0^2} = \pm 7$$

So we choose from among $\langle 7, 2n\pi\rangle$ and $\langle-7, (2n + 1)\pi\rangle$ for $\langle r, \theta\rangle$.

A quick look at $(0, 7)$ indicates that we cannot use the formulas, since $y/x$ is not defined. However, we can look directly at $(0, 7)$ in Fig. 10.11. We see that $\langle 7, \pi/2 + 2n\pi\rangle$ and $\langle-7, \pi/2 + (2n + 1)\pi\rangle$ will work. ◀

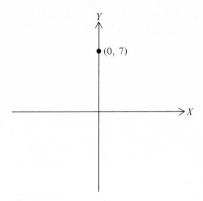

**Figure 10.11**

One use of the relationship between polar and rectangular coordinates is to change polar equations which we do not recognize into rectangular equations which we do, and vice versa.

### ▸ Illustration 10.9

Remember that in Illustration 10.4 we graphed the polar equation $r = \sin \theta$ and announced that we would later prove that the graph is a circle. Let us change from polar to rectangular coordinates to see if any light can be shed on the problem.

We know that for a point whose polar coordinates are $\langle r, \theta \rangle$ its rectangular coordinates are $(r \cos \theta, r \sin \theta)$; that is, $x = r \cos \theta$, $y = r \sin \theta$.

Our equation is $r = \sin \theta$, and we know that since $y = r \sin \theta$, $\sin \theta = y/r$. So our equation can be rewritten as

$$r = \frac{y}{r} \quad \text{or} \quad r^2 = y$$

But $r^2 = x^2 + y^2$. So by replacing $r^2$ by $x^2 + y^2$, our equation becomes

$$x^2 + y^2 = y$$

Completing the square we get

$$x^2 + y^2 - y + (\tfrac{1}{2})^2 = (\tfrac{1}{2})^2$$
$$x^2 + (y - \tfrac{1}{2})^2 = (\tfrac{1}{2})^2$$

We recognize this as a circle with center at $(0, \tfrac{1}{2})$ and radius $\tfrac{1}{2}$ (Fig. 10.12). This graph is, of course, the same as our original graph. The advantage of switching to rectangular coordinates was that we could

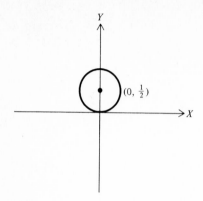

**Figure 10.12**

recognize the exact graph from the familiar form of the equation of a circle. ◄

► **Illustration 10.10**

To illustrate that it might also be useful to switch from rectangular to polar coordinates consider the following rectangular equation.

$$y = x \tan \sqrt{x^2 + y^2}$$

If you tried, you would find it quite tedious to plot several ordered pairs $(x, y)$. Instead, we change to polar coordinates as follows.

$$y = x \tan \sqrt{x^2 + y^2}$$
$$y/x = \tan \sqrt{x^2 + y^2}$$
$$\sqrt{x^2 + y^2} = \text{Tan}^{-1} y/x + n\pi$$
$$r = \theta$$

We can recognize immediately that the graph of $r = \theta$ is the spiral drawn in Illustration 10.7. ◄

Now, here are some problems for you to do.

**PROBLEMS 10.4**

1 Some points are described below by means of polar coordinates. Find the rectangular coordinates of each.

a $\langle 2, \frac{\pi}{2} \rangle$    b $\langle \sqrt{3}, -\frac{\pi}{4} \rangle$    c $\langle -7, \frac{\pi}{6} \rangle$

d $\langle \pi, \pi \rangle$    e $\langle 0, \pi \rangle$    f $\langle \pi, 0 \rangle$

**2** Find the rectangular coordinates of the points described below by means of polar coordinates.

**a** $\langle 3, 0 \rangle$     **b** $\langle -\frac{3}{2}, \frac{\pi}{6} \rangle$     **c** $\langle \sqrt{2}, \frac{\pi}{4} \rangle$

**d** $\langle 5, -\frac{\pi}{3} \rangle$     **e** $\langle 0, 0 \rangle$     **f** $\langle \frac{\pi}{2}, \frac{\pi}{2} \rangle$

**3** Find two different polar coordinate representations for each of the points whose rectangular coordinates are given below.

**a** $(1, 1)$     **b** $\left( \sqrt{2}, -\frac{\sqrt{2}}{2} \right)$     **c** $(-17, 0)$

**d** $(5, 10)$     **e** $(-5, 10)$     **f** $(0, 0)$

**4** Find two different polar coordinate representations for each of the points whose rectangular coordinates are given below.

**a** $(-3, 3)$     **b** $(12, 6)$     **c** $(\sqrt{2}, 0)$
**d** $(-2, 1)$     **e** $(e, e)$     **f** $(\pi, -2\pi)$

**5** Change each of the following polar equations to rectangular coordinates and then sketch the graph.

**a** $\theta = \frac{\pi}{4}$     **b** $r = 1$     **c** $r = \frac{3}{\cos \theta + \sin \theta}$

**6** Change each of the following polar equations to rectangular coordinates and then sketch the graph.

**a** $r = \frac{1}{\sin \theta - \cos \theta}$     **b** $r = \tan \theta \sec \theta$     **c** $r^2 = \frac{1}{2 \cos^2 \theta + \sin^2 \theta}$

**7** Change each of the following rectangular equations to polar coordinates and then graph it.

**a** $x^2 + y^2 = \text{Tan}^{-1} \frac{y}{x}$     **b** $y = x \tan \frac{1}{\sqrt{x^2 + y^2}}$

**c** $(x^2 + y^2)^3 = 2xy$     **d** $x^2 + y^2 = 25$

## 10.5  PARAMETRIC EQUATIONS

We have seen that curves can be described by rectangular equations such as $x^2 + y^2 = 25$, $y = \sin x$, and $xy = 1$, and by polar equations such as $r = \theta$, $r = 3/(\sin \theta - \cos \theta)$ and $\theta = \pi/2$. In this section curves will be expressed by means of pairs of equations, each of which gives one of the two coordinates in terms of a third variable called a *parameter*. As you will see in the following example, such parametric representations of a curve sometimes make it possible to convey more information about a situation than do the usual polar and rectangular equations.

▶ **Illustration 10.11**

Suppose an astronomer were asked to describe the path a certain comet would take across the evening sky so that local people could watch it (Fig. 10.13). He might describe the path by saying that if you stand on a certain peak and take the horizon as the $X$ axis and the tallest building downtown as the $Y$ axis, then the comet will appear to follow the path $y = 1 - x^2$, where each $x$ unit is 10 miles and each $y$ unit is 1 mile.

   This is well and good, but what if the people also wanted to know when to look where for the comet? The astronomer could reply that if $t$ stands for the time after midnight in hours, then for $0 \leq t \leq 2$, the comet's coordinates would be given by $x = t - 1$, $y = 2t - t^2$. Then, for example, if a citizen wanted to know where the comet would be at 12:30 (that is, $t = \frac{1}{2}$), simply plugging $t = \frac{1}{2}$ into $(t - 1, 2t - t^2)$ would provide the information that the comet would be at $(-\frac{1}{2}, \frac{3}{4})$. By giving the rectangular coordinates of the comet in terms of the parameter $t$ the astronomer would be able to describe not only the path of the comet, but also where the comet was at each moment in time. ◀

▶ **Illustration 10.12**

Graph the curve whose polar coordinates are given parametrically as follows.

$$r = t^2$$
$$\theta = 2t \qquad \text{for } t \geq 0$$

One way to graph the curve is to plot points. (Note that only a few sample points from the table are pictured.) See Fig. 10.14.

| $t$ | 0 | $\dfrac{\pi}{12}$ | $\dfrac{\pi}{8}$ | $\dfrac{\pi}{4}$ | $\dfrac{\pi}{2}$ | $\pi$ | $3\dfrac{\pi}{2}$ | $2\pi$ |
|---|---|---|---|---|---|---|---|---|
| $\theta$ | 0 | $\dfrac{\pi}{6}$ | $\dfrac{\pi}{4}$ | $\dfrac{\pi}{2}$ | $\pi$ | $2\pi$ | $3\pi$ | $4\pi$ |
| $r$ | 0 | $\dfrac{\pi^2}{144}$ | $\dfrac{\pi^2}{64}$ | $\dfrac{\pi^2}{16}$ | $\dfrac{\pi^2}{4}$ | $\pi^2$ | $\dfrac{9}{4}\pi^2$ | $4\pi^2$ |

Another way to graph this curve would have been to notice that

$$r = t^2 \qquad \text{and} \qquad \theta = 2t$$

Solving $\theta = 2t$ for $t$, we get $\qquad t = \dfrac{\theta}{2}$

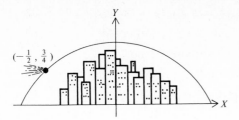

**Figure 10.13**

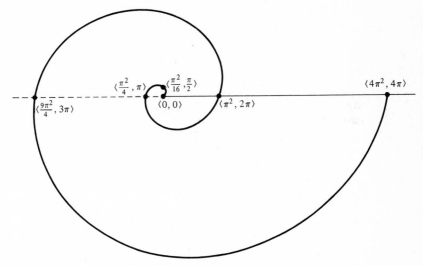

**Figure 10.14**

Substituting into $r = t^2$, we get

$$r = \left(\frac{\theta}{2}\right)^2$$

This is pretty easily seen to be a spiral of the kind we have drawn. ◀

◀ **Illustration 10.13**

Graph
$$x = 3 \sin t$$
$$y = 3 \cos t \qquad 0 \leq t \leq \pi$$

Again we could plot pairs $(x, y)$ for several different values of $t$. However, the work involved is considerably reduced if we notice that

$$x^2 + y^2 = (3 \sin t)^2 + (3 \cos t)^2$$
$$= 9 \sin^2 t + 9 \cos^2 t$$
$$= 9(\sin^2 t + \cos^2 t)$$
$$= 9$$

We can quickly recognize that $x^2 + y^2 = 9$ is the equation of the circle with radius 3 and center at $(0, 0)$ (Fig. 10.15). Then we observe that for $0 \leq t \leq \pi$ we get values of $x$ ranging from 0 to 3 back to 0, and values of $y$ ranging from 3 to 0 on to $-3$. So our graph is the semicircle illustrated in Fig. 10.15. It can also be noted that as $t$ "goes" from 0 to $\pi$, the point on the graph "goes" from $(0, 3)$ to $(0, -3)$. ◀

▶ **Illustration 10.14**

One more example of eliminating the parameter. Suppose that a curve is described parametrically as follows.

$$x = 29t + 3$$
$$y = \frac{t - 1}{2}$$

Solving for $t$ we get

$$t = \frac{x - 3}{29}$$

so

$$y = \frac{\left(\dfrac{x - 3}{29}\right) - 1}{2} = \frac{x - 32}{58}$$

(See Fig. 10.16.) ◀

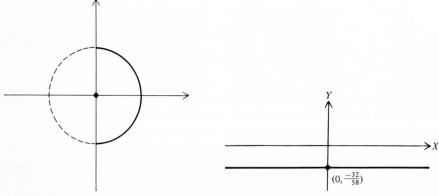

**Figure 10.15**                    **Figure 10.16**

## PROBLEMS 10.5

1   Locate the comet in Illustration 10.11 at 12:45, 1:00, 1:30, 1:50, and 2:00. What was the apparent horizontal velocity of the comet in miles per hour?

2   Graph the curve described by the following parametric equations.

$$x = 3 \cos t$$
$$y = 2 \sin t \qquad \frac{\pi}{2} \leq t \leq \frac{3\pi}{2}$$

3   Graph the curve described by the following parametric equations.

$$\theta = 3t$$
$$r = \sin 6t \qquad 0 \leq t \leq \frac{2\pi}{3}$$

4   Graph the curve described by the following parametric equations.

$$x = 2t + 7$$
$$y = \frac{t-1}{3} \qquad -5 \leq t \leq 4$$

5   Suppose that a cyclotron is circular as shown in Fig. 10.17. Make up a story like that in Illustration 10.11 about locating a particle in the cyclotron using parametric equations.

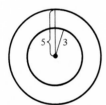

**Figure 10.17**

6   Make up a complicated-looking parametric representation for a simple curve. Then see if a classmate can eliminate the parameter and graph the curve.

# GRAPHING
# IN SPACE

## 11.1   LOCATING POINTS

You have already had experience with locating points on a flat surface
(a plane) using ordered pairs of numbers. For example, (1, 1), (0, 0),
(−3, 5), (5, −3), and (−1, −1) are located in Fig. 11.1. Each of the pairs
of numbers uniquely locates a point relative to the intersecting lines
(axes) on which unit distances are marked.

Many of the interesting shapes in the world are not flat and cannot
be represented in terms of points on a plane. It is therefore very natural
to extend the above point representations to space by representing
points relative to three perpendicular intersecting axes using ordered
triples of numbers.

We start with three mutually perpendicular coordinate axes, with
unit lengths indicated as in Fig. 11.2.

The usual way of labeling the three axes is as X axis, Y axis, and
Z axis. While there are three possible ways of assigning these labels to
the axes, two ways are most common and are pictured in Figs. 11.3 and

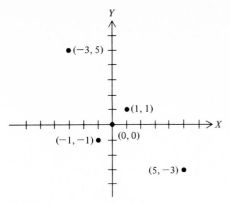

Figure 11.1

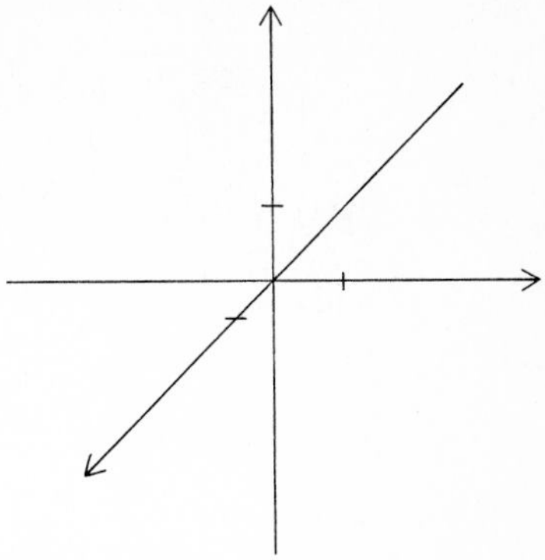

Figure 11.2

11.4. The axes labeled as in Fig. 11.3 form what is called a left-hand system. Labeled as in Fig. 11.4 the axes form a right-hand system, which is the kind that will be used in this chapter. Each pair of coordinate axes determines a plane which we call a *coordinate plane.* These three coordinate planes are called the XY plane, the XZ plane and the YZ plane and are shaded in Figs. 11.5 to 11.7.

Each point in space can be located by specifying three numbers called *coordinates* which indicate the distances of the point from the three coordinate planes. The coordinates are listed in the order (x, y, z). The x coordinate designates the distance of the point from the YZ plane, the y coordinate designates the distance from the XZ plane, and the z coordinate designates the distance from the XY plane. So, for example, the ordered triple (1, 2, 3) designates the point which has distance 1 from the YZ plane, distance 2 from the XZ plane, and dis-

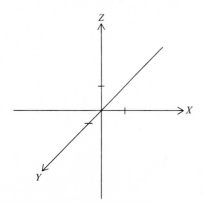

Figure 11.3

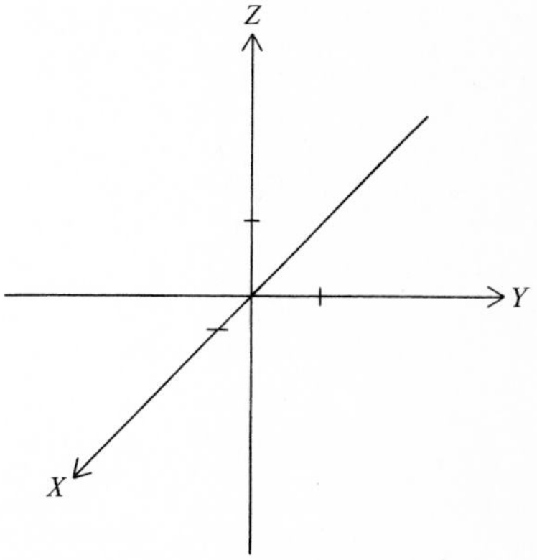

Figure 11.4

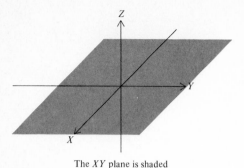

The *XY* plane is shaded

**Figure 11.5**

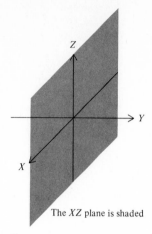

The *XZ* plane is shaded

**Figure 11.6**

tance 3 from the *XY* plane (Fig. 11.8). The dotted lines indicate that we located the point by "traveling" 1 unit out the *X*-axis, then 2 units parallel to the *Y* axis, and then 3 units parallel to the *Z* axis. Further illustrations follow.

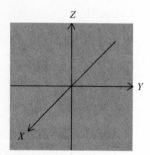

The *YZ* plane is shaded

**Figure 11.7**

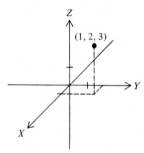

**Figure 11.8**

## ◗ Illustration 11.1

The following points are located in Fig. 11.9: $(0, 0, 0)$, $(1, 1, 1)$, $(-1, 1, 1)$, $(1, -1, 1)$, $(1, 1, -1)$, $(-1, -1, 1)$, $(1, -1, -1)$, $(-1, 1, -1)$, and $(-1, -1, -1)$. Notice that some guidelines have been dotted in to help in visualizing the location of the points. Notice also that the first number in the ordered triple $(x, y, z)$ locates the point relative to the axis which comes "toward" you. The second coordinate locates the point relative to the axis which runs "east-west," and the third number tells where the point is relative to the "north-south" axis. ◗

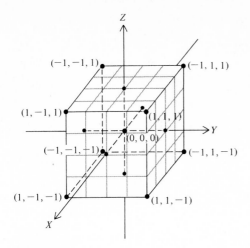

**Figure 11.9**

### ◆ Illustration 11.2

Two points are indicated in Fig. 11.10, and then their coordinates are given.

Any point can be located by an ordered triple, and any ordered triple locates a point. ◆

That is all that there is to it. In the forthcoming exercises you should take every opportunity to draw pictures, since the ability to visualize and draw space figures requires practice for many people and can prove to be a most useful skill.

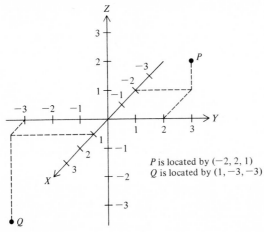

$P$ is located by $(-2, 2, 1)$
$Q$ is located by $(1, -3, -3)$

**Figure 11.10**

## PROBLEMS 11.1

**1** Draw three perpendicular axes, mark a unit length on each axis, and then locate the points represented by the following ordered triples relative to the axes. (Dot in any guidelines that are helpful.)

    **a** $(-2, -2, -2)$      **b** $(\frac{1}{2}, -1, 3)$      **c** $(\sqrt{2}, 2, 2)$      **d** $(0, 0, 4)$

**2** As in Prob. 1, locate each of the following ordered triples relative to three perpendicular axes.

    **a** $(0, 3, 0)$      **b** $(0, 0, -4)$      **c** $(0, 1, 2)$
    **d** $(2, 1, 0)$      **e** $(\sqrt{2}, 0, \sqrt{2})$

**3** Find the ordered triples that represent the points $P$ and $Q$ pictured in Fig. 11.11.

**4** Find the ordered triples that represent the points $P$ and $Q$ pictured in Fig. 11.12.

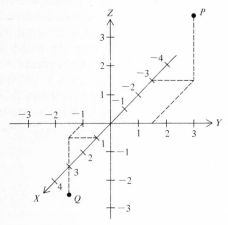

**Figure 11.11**

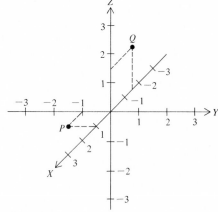

**Figure 11.12**

## 11.2 DISTANCE AND SPHERES

In this section we want to define what we mean by distance in space. To do this we will first define distance by means of a formula, and then we will indicate why we chose that particular definition. Finally we will apply the formula for distance to the problem of finding equations of spheres.

### DEFINITION 11.1

The distance between the points $(x, y, z)$ and $(a, b, c)$ is given by

$$\sqrt{(x - a)^2 + (y - b)^2 + (z - c)^2}$$

To get a "feeling" for the definition, look at Fig. 11.13. In order for our definition to be consistent with the pythagorean theorem from plane geometry, the distance $d$ should satisfy the following equation

$$d^2 = S_1{}^2 + S_2{}^2$$

That is,

$$d = \sqrt{S_1{}^2 + S_2{}^2} \tag{1}$$

Now $S_2{}^2 = (z - c)^2$. Check and be sure. Again, using the pythagorean theorem to find $S_1{}^2$,

$$S_1{}^2 = S_3{}^2 + S_4{}^2 = (x - a)^2 + (y - b)^2$$

So, substituting $S_1{}^2$ and $S_2{}^2$ into formula (1) we get

$$d = \sqrt{(x - a)^2 + (y - b)^2 + (z - c)^2} \tag{1'}$$

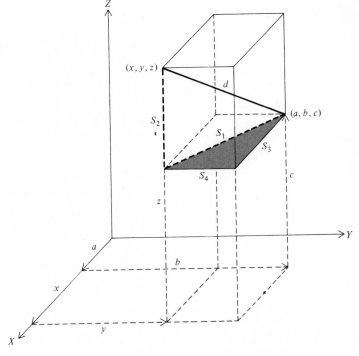

**Figure 11.13**

♦ **Illustration 11.3**

Find the distance between $(1, 2, 3)$ and $(\frac{1}{2}, -\frac{3}{4}, -2)$.

$$d = \sqrt{(1 - \tfrac{1}{2})^2 + [2 - (-\tfrac{3}{4})]^2 + [3 - (-2)]^2}$$
$$= \sqrt{(\tfrac{1}{2})^2 + (\tfrac{11}{4})^2 + 5^2}$$
$$= \sqrt{\tfrac{1}{4} + \tfrac{121}{16} + 25}$$
$$= \sqrt{\tfrac{525}{16}} \quad \blacktriangleleft$$

▶ **Illustration 11.4**

Find the distance between $(0, 1, 1)$ and $(3, 2, 1)$.

$$d = \sqrt{(0 - 3)^2 + (1 - 2)^2 + (1 - 1)^2}$$
$$= \sqrt{9 + 1 + 0}$$
$$= \sqrt{10} \quad \blacktriangleleft$$

Before discussing spheres, recall what it means for a shape to be the graph of an equation.

**DEFINITION 11.2**

A shape is the graph of an equation if

    **a**   The coordinates of each point on the shape satisfy the equation.
    **b**   Each set of real numbers which satisfy the equation are the coordinates of a point on the shape  ◀

You recall from your previous graphing work, for example, that $x^2 + y^2 = 4$ is the equation of a circle in the plane (Fig. 11.14), since

    **1**   Any point $(x, y)$ on the circle is (by definition of a circle) at a distance of 2 from $(0, 0)$. So,

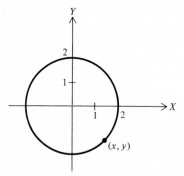

**Figure 11.14**

So substituting into the original equation we get

$$(x - 1)^2 - 1 + (y - 2)^2 - 4 + (z - 3)^2 - 9 + 5 = 0$$

or

$$(x - 1)^2 + (y - 2)^2 + (z - 3)^2 = 9$$

We can recognize this latter equation as that of a sphere with center $(1, 2, 3)$ and radius 3.  ◀

## PROBLEMS 11.2

1   Find the distances between the following pairs of points.

    **a**   $(0, 0, 0)$ and $(1, 1, 1)$    **b**   $(\sqrt{2}, 0, \sqrt{2})$ and $(0, -\sqrt{5}, 0)$

    **c**   $(\frac{3}{2}, 1, 2)$ and $(1, \frac{5}{3}, \frac{1}{4})$    **d**   $(\pi, -2, 1)$ and $(3, 2, 1)$

2   Find the distances between the following pairs of points.

    **a**   $(0, 0, 0)$ and $(1, -1, -1)$    **b**   $(\frac{1}{2}, \frac{1}{3}, \frac{1}{4})$ and $(\frac{1}{4}, \frac{1}{3}, \frac{1}{2})$

    **c**   $(-2, 1, -2)$ and $(-3, 2, 3)$    **d**   $(\sqrt{2}/2, \frac{1}{2}, \frac{3}{2})$ and $(\sqrt{3}/3, \frac{1}{3}, \frac{4}{3})$

3   Sketch and find an equation for each of the spheres described below.
    **a**   The sphere with center at the origin and radius 4
    **b**   The sphere with center at $(-1, -1, -1)$ and radius $\frac{1}{2}$
    **c**   The sphere which has center at $(1, \sqrt{2}, 3)$ and is tangent to the YZ plane

4   Sketch and find an equation for each of the spheres described below.
    **a**   The sphere with center $(0, 1, 0)$ and radius 1
    **b**   The sphere with center $(-\frac{1}{2}, -\frac{1}{3}, 1)$ and radius $\frac{3}{4}$
    **c**   The sphere with center $(0, 0, 4)$ which is tangent to the XY plane

5   Describe and sketch each of the following.
    **a**   The graph of $x^2 + y^2 + z^2 = \frac{1}{4}$
    **b**   The graph of $(x + 1)^2 + (y - 4)^2 + (z + 2)^2 = 9$
    **c**   The graph of $x^2 - x + y^2 + 2y + z^2 - \frac{7}{4} = 0$

6   Describe and sketch each of the following.
    **a**   The graph of $(x + 1)^2 + (y + 1)^2 + (z + 1)^2 = 1$
    **b**   The graph of $x^2 + y^2 + (z - \frac{1}{2})^2 = \frac{1}{4}$
    **c**   The graph of $x^2 + \frac{2}{3}x + y^2 + z^2 - \frac{17}{9} = 0$

7   Find a general form of the equation of the sphere which is tangent to all three coordinate planes and whose center has all three coordinates positive.

8   Describe the graph of

$$x^2 + y^2 + z^2 + Ax + By + Cz + D = 0$$

for all possible values of $A$, $B$, $C$, and $D$.

## 11.3    PLANES

In this section you will have an opportunity to gain skill with graphing the equations of planes.  Certain of the simplest planes discussed will be useful in your graphing work in the next two sections.

For an analog let us look back at graphing equations in x and y in the plane.  You will recall that the graph of $x = 1$ is the set of all points whose x coordinates are 1, i.e., points of the form $(1, y)$ for any y (Fig. 11.16).  Also the graph of $y = -\frac{3}{2}$ consists of points of the form $(x, -\frac{3}{2})$.

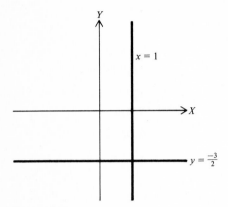

**Figure 11.16**

In space an analogous thing happens.  The graph of $x = 1$ is the set of all points of the form $(1, y, z)$ for any y and z (Fig. 11.17).  This graph

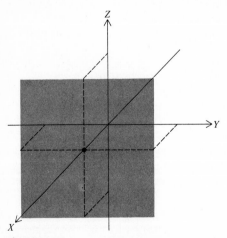

**Figure 11.17**

is a plane which is parallel to the YZ plane at a distance 1 from it. The X axis is perpendicular to the graph of x = 1, and it intersects it in (1, 0, 0).

In the same way $y = -\frac{3}{2}$ and z = 2 are planes parallel to the XZ and XY planes, respectively (Fig. 11.18).

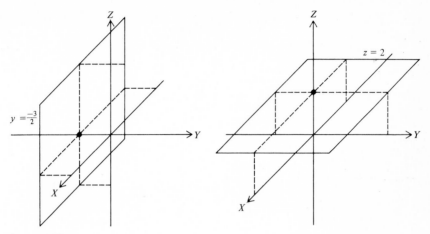

**Figure 11.18**

To cement your understanding of this material you should stop right now and sketch the graphs of $x = -\frac{1}{2}$, y = 1, and z = -3 in space.

We have seen how to graph equations in space that only restrict one of the variables x, y, and z. Now we consider equations which restrict two of the variables. Fortunately, we can again work with an analog from graphing in the plane. In earlier chapters you found that the graph of x + y = 1 is the straight line pictured in Fig. 11.19. In three dimensions, the graph of the equation x + y = 1 is the set of all points

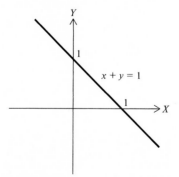

**Figure 11.19**

$(x, y, z)$ such that $x + y = 1$, and $z$ can have any value (Fig 11.20a). The graph is the plane pictured which is perpendicular to the $XY$ plane. One way of visualizing the graph is by taking cross sections parallel to the $XY$ plane. That is, intersect it with planes of the form $z = k$. Each such plane intersects the graph in the line $x + y = 1$ (Fig. 11.20b).

For another example, the graph of $y = z$ is perpendicular to the $YZ$ plane and intersects each plane of the form $x = k$ in the straight line $y = z$. You should stop and sketch the graph.

Summarizing what we have done to date, we have the following rule.

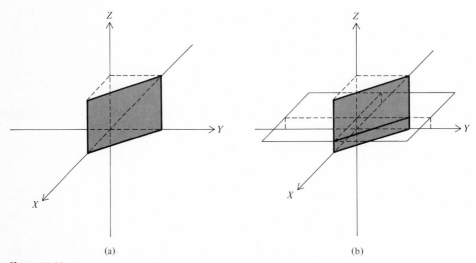

(a)                                (b)

**Figure 11.20**

### Graphing Planes

**1** The graph of any equation of the form $x = k$ is a plane which is parallel to the $YZ$ plane and which is perpendicular to the $X$ axis, intersecting it in the point $(k, 0, 0)$.

**2** The graph of any equation of the form $y = k$ is a plane which is parallel to the $XZ$ plane and which is perpendicular to the $Y$ axis, intersecting it in the point $(0, k, 0)$.

**3** The graph of any equation of the form $z = k$ is a plane parallel to the $XY$ plane and which is perpendicular to the $Z$ axis, intersecting it in the point $(0, 0, k)$.

**4** The graph of any equation of the form $ax + by = d$ is a plane which is perpendicular to the $XY$ plane and intersects the graph of $z = k$ in the straight line whose two-dimensional equation is $ax + by = d$.

**5** The graph of any equation of the form $by + cz = d$ is a plane which is perpendicular to the *YZ* plane and intersects the graph of $x = k$ in the straight line whose two-dimensional equation is $by + cz = d$.

**6**   The graph of any equation of the form $ax + cz = d$ is a plane which is perpendicular to the *XZ* plane and intersects the graph of $y = k$ in a straight line whose two-dimensional equation is $ax + cz = d$.

Since the main purpose of this chapter is to give you some experience with three-dimensional graphing, we will state the next theorems concerning the general equation for planes without developing the machinery needed to prove it. We will then provide you with some examples and some exercises.

---

**THEOREM**   *The general equation of a plane is*
**11.2**

$$ax + by + cz = d$$

*That is, any plane has an equation of this form, and the graph of any equation of this form is a plane (providing at least one of a, b, and c is not 0).*

---

▶ **Illustration 11.7**

Graph $x + y + z = 1$.

First, from Theorem 11.2 we can conclude that the graph is a plane. Then to determine the orientation of the plane, check its intersections with the three coordinate planes. To find its intersection with the *XY* plane, we set $z = 0$. Then $x + v = 1$. Similarly for the *YZ* plane, $x = 0$ so $y + z = 1$; and for the *XZ* plane, $y = 0$ so $x + z = 1$ (Fig. 11.21). Thus

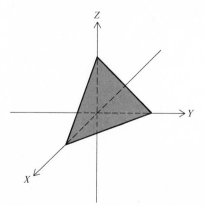

**Figure 11.21**

visualizing the plane in the first octant (i.e., where $x \geq 0$, $y \geq 0$, $z \geq 0$) we get a pretty clear picture of the plane which actually continues indefinitely in all directions. ◀

▶ **Illustration 11.8**

We can graph $2x - y + z = 3$ in a similar manner (Fig. 11.22).

If $x = 0$,        $-y + z = 3$
If $y = 0$,        $2x + z = 3$
If $z = 0$,        $2x - y = 3$

Combining these we get Fig. 11.23. ◀

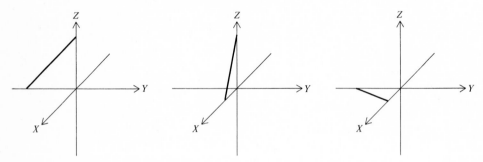

**Figure 11.22**

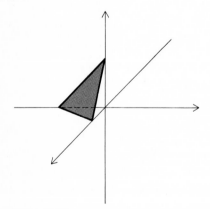

**Figure 11.23**

▶ **Illustration 11.9**

Find an equation of the plane which passes through the points $(1, 0, 2)$,

$(1, -1, -1)$, and $(\frac{1}{2}, 1, -\frac{1}{2})$. We know that the equation has the form

$$ax + by + cz = d$$

We also know that the coordinates of each of the given points must satisfy the equation. So, substituting the coordinates of the points into the equation we get three equations in the four unknowns $a, b, c,$ and $d$.

$$a(1) + b(0) + c(2) = d \qquad \text{or} \qquad a + 2c = d \qquad (1)$$
$$a(1) + b(-1) + c(-1) = d \qquad \text{or} \qquad a - b - c = d \qquad (2)$$
$$a\left(\frac{1}{2}\right) + b(1) + c\left(-\frac{1}{2}\right) = d \qquad \text{or} \qquad \frac{a}{2} + b - \frac{c}{2} = d \qquad (3)$$

A standard trick here is to express each of $a, b,$ and $c$ in terms of $d$. (If $d$ had been 0, we would have had three equations and three unknowns.)

Adding (2) and (3) we get

$$a - b - c + \frac{a}{2} + b - \frac{c}{2} = d + d \qquad \text{or} \qquad \frac{3a}{2} - \frac{3c}{2} = 2d \qquad (4)$$

Multiplying (1) by $-\frac{3}{2}$ and adding it to (4) we get

$$\frac{-3}{2}a - 3c + \frac{3}{2}a - \frac{3c}{2} = -\frac{3}{2}d + 2d \qquad -\frac{9}{2}c = \frac{1}{2}d \qquad \text{or} \qquad c = -\frac{1}{9}d \quad (5)$$

Substituting (5) into (1) we get

$$a + 2\left(-\tfrac{1}{9}d\right) = d \qquad \text{or} \qquad a = \tfrac{11}{9}d \qquad (6)$$

Finally, substituting (5) and (6) into (2) we get

$$\tfrac{11}{9}d - b - \left(-\tfrac{1}{9}d\right) = d \qquad \text{or} \qquad b = \tfrac{3}{9}d$$

We then have that

$$a = \tfrac{11}{9}d$$
$$b = \tfrac{3}{9}d$$
$$c = -\tfrac{1}{9}d$$

From this we can get our desired equation by letting $d$ have any value besides 0. For example, if we let $d = 9$, we get

$$11x + 3y - z = 9$$

If we had let $d = 18$, we would have gotten $22x + 6y - 2z = 18$. Since the second equation is the result of multiplying the first by 2, the two equations clearly have the same graph. You may want to check one of these equations to see if the three points specified in this illustration lie on its graph. ◂

**PROBLEMS 11.3**

1 Sketch the graph of each of the following equations in three dimensions.

    **a** $x = -2$     **b** $y = \frac{3}{2}$     **c** $z = -\sqrt{2}$

2 Sketch the graph of each of the following equations in three dimensions.

    **a** $x = \pi$     **b** $y = -1$     **c** $z = 3$

3 Sketch the graph of each of the following equations in three dimensions.

    **a** $x - y = 1$     **b** $y = -z$     **c** $2x + z = 4$

4 Sketch the graph of each of the following equations in three dimensions.

    **a** $-x + 2y = 7$     **b** $y + z - 2 = 0$     **c** $\frac{1}{2}x + z = 3$

5 Sketch the graph of each of the following equations.

    **a** $-x + y - 2z = 1$

    **b** $x + \dfrac{y}{3} - 3z = 5$

    **c** $\sqrt{2}x + \sqrt{3}y + \sqrt{3}z = \sqrt{4}$

6 Sketch the graph of each of the following equations.

    **a** $x + y + z = -1$     **b** $2x - y + 4z = 3$     **c** $8x + 7y + 2z = 1$

7 Sketch and find the equation of the plane which is perpendicular to the XY plane and intersects the plane $z = 2$ in the line $y = x$.

8 Sketch and find the equation of the plane which intersects the XY plane in the line $3x + 2y = 1$, the XZ plane in the line $3x + 4z = 1$, and the YZ plane in the line $2y + 4z = 1$.

9 Sketch and find the equation of the plane which does not intersect the XY plane and intersects the Z axis in the point $(0, 0, \sqrt{2})$.

10 Sketch and find the equation of the plane which contains the points $(0, 1, 0)$, $(1, 0, 0)$, and $(0, 0, 1)$.

11 Sketch and find the equation of the plane which contains the points $(1, 0, 0)$, $(0, 1, 0)$, $(0, 0, 2)$.

12 Find an equation of the plane which contains the points $(1, 0, -1)$, $(0, 1, 2)$, $(1, 1, 0)$, and $(0, 0, 1)$.

**11.4  GRAPHS OF CYLINDERS AND CURVES**

Continuing the study of graphing in three dimensions, we will look at some equations whose graphs can be drawn using what we know about graphing in two dimensions. The key feature of these equations will be that at least one of the three variables is missing.  In Sec. 11.3 you studied equations like $x = 3$ and $x + y = 1$.  You will recall that the graph of $x = 3$ is a plane which is parallel to the YZ plane and therefore parallel to both the Y axis and the Z axis.  Furthermore, the graph of $x + y = 1$ is a plane which is perpendicular to the XY plane and there-fore parallel to the Z axis.  Let us look at a couple more examples before making generalizations

### ♦ Illustration 11.10

Graph $y^2 + z^2 = 4$.

In two dimensions (e.g., in the YZ plane), the graph of $y^2 + z^2 = 4$ is a circle with center at the origin and radius 2 (Fig. 11.24).  Since x does not appear in the equation, any point $(x, y, z)$ will lie on the graph as long as y and z satisfy the equation.  For example, $(x, 0, 2)$ satisfies the equation for any value of x.  The points $(x, 0, 2)$ lie on a straight line which is parallel to the X axis and which passes through $(0, 0, 2)$ (Fig. 11.25a).  Each point on the circle "generates" such a straight line.  So the resulting graph is a circular cylinder whose sides are parallel to the X axis and which intersects the YZ plane in the circle $y^2 + z^2 = 1$ (Fig. 11.25b).  The circular cylinder intersects each plane which is perpen-dicular to the X axis (i.e., graph of $x = k$) in a circle (Fig. 11.25c).  ♦

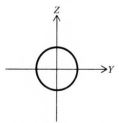

**Figure 11.24**

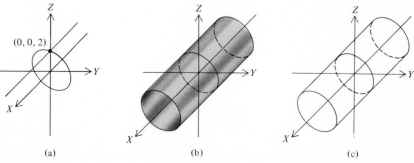

(a)          (b)          (c)

**Figure 11.25**

♦ **Illustration 11.11**

Graph $y = x^2$ in three dimensions.

You will recall that in two dimensions the graph of $y = x^2$ is the parabola pictured in Fig. 11.26. This time, since z is missing, (x, y, z) is on the graph for any value of z as long as $y = x^2$. So each point on the parabola generates a line parallel to the Z axis which is in the graph. In Fig. 11.27a you can see the line generated by (2, 4, 0).

So the resulting graph is a parabolic cylinder the sides of which are parallel to the Z axis (Fig. 11.27b). In this case cross sections which result from intersecting $y = x^2$ with planes perpendicular to the Z axis (i.e., graphs of $z = k$) are parabolas. ◄

We are now ready for a generalization.

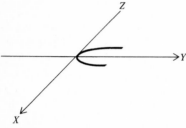

**Figure 11.26**

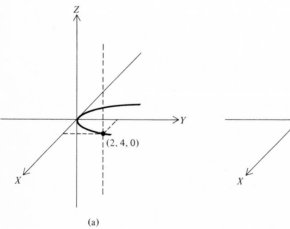

(a)

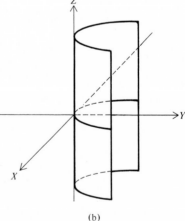

(b)

**Figure 11.27**

### Cylinders

The graph in three dimensions of an equation in two variables (i.e., one variable missing) is the *cylinder* generated by the two-dimensional graph of the equation. The sides of the cylinder are parallel to the axis of the missing variable. The cross section resulting from intersecting the cylinder with a plane perpendicular to the axis of the missing variable is the curve which is the two-dimensional graph of the equation.

The illustrations and the generalization raise the other topic of this section, namely, curves. In the examples we saw that the intersection of the cylinders and certain planes were curves. We will pursue this idea further with some more examples.

### ▶ Illustration 11.12

Let us investigate the intersections of the graphs of $x^2 + y^2 + z^2 = 4$ and $z = 1$. Algebraically, we can see that substituting $z = 1$ into $x^2 + y^2 + z^2 = 4$ gives $x^2 + y^2 = 3$, which is the equation of a circle with radius $\sqrt{3}$. Graphically, this makes sense because the graph of $x^2 + y^2 + z^2 = 4$ is a sphere and the graph of $z = 1$ is a plane perpendicular to the z axis (Fig. 11.28).

So we can say that the circle pictured in Fig. 11.28 is the set of points which simultaneously satisfy the two equations

$$x^2 + y^2 + z^2 = 4$$

and $$z = 1 \quad ◀$$

### ▶ Illustration 11.13

Graph the simultaneous solutions of

$$x^2 + y^2 + z^2 = 4$$

and $$y = x$$

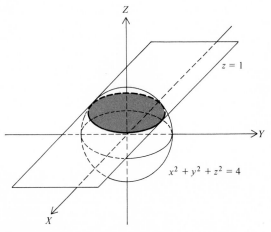

**Figure 11.28**

The graphs of the separate equations are, respectively, the sphere with center at the origin and radius 2 and the plane (cylinder) generated by the line $y = x$. Geometric intuition suggests that the plane $y = x$ will "slice" the sphere in half, yielding a circle of radius 2 with center at the origin lying in the plane $y = x$ (Fig. 11.29).

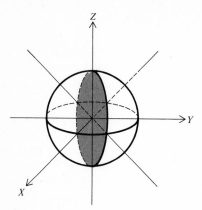

**Figure 11.29**

There might be a source of confusion here, though. If you substitute $y = x$ into $x^2 + y^2 + z^2 = 4$, you get, for example, $2y^2 + z^2 = 4$. This is the equation of an ellipse in the $YZ$ plane. This ellipse is the "projection" of the circular intersection of the plane and the sphere onto the $YZ$ plane, as pictured in Fig. 11.30. It is as though you were casting

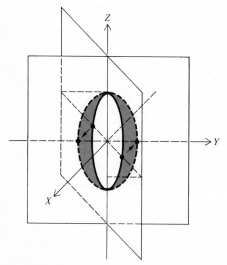

**Figure 11.30**

a shadow of the circle on the YZ plane using a light that is parallel to the X axis. ◀

◆ **Illustration 11.14**

Graph the simultaneous solutions of $x^2 + y^2 = 9$ and $y = 2$ (Fig. 11.31).
    Again, using our intuition, we can guess that the plane $y = 2$ will intersect the cylinder in two parallel lines. The lines are parallel to the Z axis. Clearly, since they lie in the plane $y = 2$, every $y$ coordinate on each line is 2. Substituting $y = 2$ into $x^2 + y^2 = 9$, we get $x^2 = 5$, or $x = \sqrt{5}$ and $x = -\sqrt{5}$. So the lines consist of points $(\sqrt{5}, 2, z)$ and and $(-\sqrt{5}, 2, z)$ for all values of z. ◀

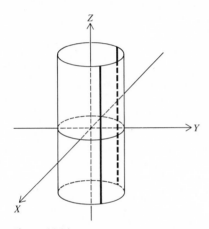

**Figure 11.31**

    In working the following exercises use your geometric intuition freely and gain as much practice with sketching as possible.

## PROBLEMS 11.4

**1** Sketch the graph of each of the following equations in three dimensions.

    **a**  $2x + y = 1$    **b**  $y = 1/z$    **c**  $x^2 + 3z^2 = 2$

**2** Sketch the graph of each of the following equations in three dimensions.

    **a**  $y - z = 3$    **b**  $x^2 - y^2 = 1$    **c**  $(x - 1)^2 + (y - 2)^2 = 4$

**3** Sketch a graph of the simultaneous solutions of $z = 1/y$ and $x = 3$ in three dimensions.

**4**   Sketch a graph of the simultaneous solutions of $x^2 + z^2 = 4$ and $z = x$.

**5**   Sketch the intersection of the graphs of the following two equations: $z = \sin y$ and $x = -2$.

**6**   Sketch the intersection of the graphs of the following two equations: $x = \log_2 y$ and $x^2 + y^2 + z^2 = 4$.

## 11.5    CROSS SECTIONS AS A TOOL IN GRAPHING

As you have seen, there is a general equation for planes, so that in one sense one can completely analyze the situation concerning the equations of planes and their graphs. One can do the same thing with the general second-degree equation, i.e., the equation

$$ax^2 + by^2 + cz^2 + dxy + exz + fyz + gx + hy + iz + j = 0$$

Spheres, of course, are graphs of special cases of this equation. One can analyze all the possible graphs that can result from various choices of the coefficients $a$, $b$, $c$, $d$, $e$, $f$, $g$, $h$, $i$, and $j$. Instead, we will describe a general approach to graphing that will be helpful in graphing many different kinds of equations, including all second-degree equations. The general approach is a natural sequel to the previous sections, because it involves determining what a surface looks like by finding out what cross sections of the surface look like.

▶ **Illustration 11.15**

To graph $x^2 + 2y^2 + z^2 = 3$ we will intersect it with planes of the form $x = k$, $y = k$, and $z = k$ for various values of $k$. For $x = k$, the equation becomes $2y^2 + z^2 = 3 - k^2$. You may recall that for $3 - k^2 > 0$ these are ellipses. Now $3 - k^2 > 0$ if $-\sqrt{3} < k < \sqrt{3}$. All of this says that the planes $x = k$ (perpendicular to the $X$ axis) intersect the shape in ellipses for $-\sqrt{3} < k < \sqrt{3}$. Also notice that as $k$ nears $-\sqrt{3}$ and $\sqrt{3}$, the size of the ellipses gets smaller (Fig. 11.32a).

For $y = k$, $x^2 + z^2 = (3 - 2k^2)$. These are circles as long as $3 - 2k^2 > 0$, that is, as long as $-\sqrt{\frac{3}{2}} < k < \sqrt{\frac{3}{2}}$ (Fig. 11.32b). For $z = k$, $x^2 + 2y^2 = 3 - k^2$. This is like the case of $x = k$, only the cross sections are perpendicular to the $Z$ axis (Fig. 11.32c).

Let us put together the information we now have about the surface. We know that the figure is confined between $\pm\sqrt{3}$ on the $X$ and $Z$ axes, and $\pm\sqrt{\frac{3}{2}}$ on the $Y$ axis. It has circular cross sections perpendicular to the $Y$ axis and elliptic cross sections perpendicular to the $X$ and $Z$ axes. The surface has the shape of a "squashed" football. It is one of a family of surfaces called *ellipsoids* (Fig. 11.32d).  ◀

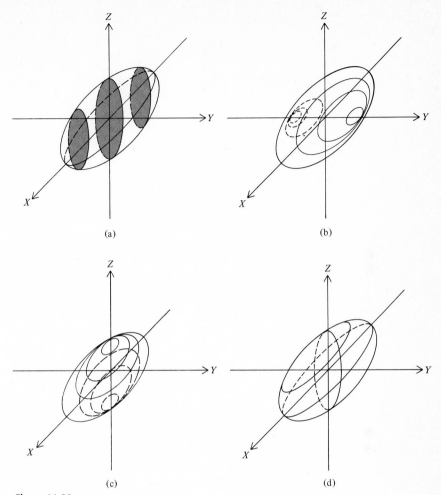

(a)

(b)

(c)

(d)

**Figure 11.32**

◆ **Illustration 11.16**

Graph $x^2 + y^2 = z$.

Since $x^2 + y^2 \geq 0$, you know that $z \geq 0$. So the entire graph lies above the $XY$ plane. Moreover, for $z = k > 0$, $x^2 + y^2 = k$ gives a circle; and the circle gets larger as $z$ does (Fig. 11.33a).

Now for $y = k$ and $x = k$, you get $z = x^2 + k^2$ and $z = y^2 + k^2$. These are parabolas with the same shape which have higher minimum points for larger values of $k^2$ (Fig. 11.33b and c). Again our composite is a little tricky, but it can be visualized as a round-nosed bullet resting on the $XY$ plane (Fig. 11.33d). ◀

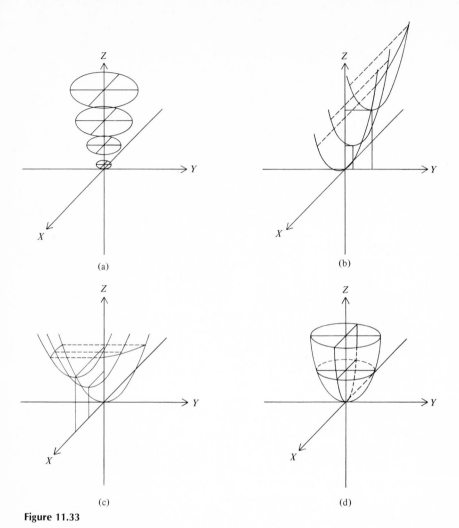

(a)

(b)

(c)

(d)

**Figure 11.33**

## PROBLEMS 11.5

**1**  Sketch the graph of $3x^2 + 2y^2 + z^2 = 4$.
**2**  Sketch the graph of $x^2 + y^2 = z^2$.
**3**  Sketch the graph of $y^2 + z^2 = -x$.
**4**  Sketch the graph of $z = 1/(x + y)$.
**5**  Sketch the graph of $x^2 + y^2 = (\sin z)^2$.
**6**  Sketch the graph of $z = \sin (x + y)$.
**7**  Sketch the graph of $x^2 + 4z^2 + 4y^2 = 9$.
**8**  Sketch the graph of $z = x^2 - y^2$.

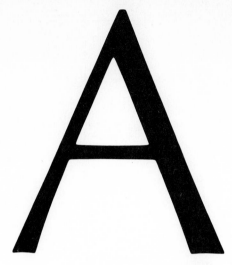

# SYNTHETIC DIVISION

Synthetic division is a shorthand technique for dividing polynomials by polynomials of the form $x - r$. Essentially what is done is to leave out all the variables involved and just copy down the coefficients. Consider the following illustration

$$
\begin{array}{r}
2x^2 + \phantom{0}2x + 19 \\
x - 4 \overline{)\, 2x^3 - 6x^2 + 11x - \phantom{0}6} \\
\underline{2x^3 - 8x^2} \phantom{+ 11x - 6} \\
2x^2 + 11x \phantom{- 6} \\
\underline{2x^2 - \phantom{0}8x} \phantom{- 6} \\
19x - \phantom{0}6 \\
\underline{19x - 76} \\
+ 70
\end{array}
$$

This process can be shortened to

$$
\begin{array}{r|rrrr}
-4 & 2 & -6 & +11 & -6 \\
& & -8 & -8 & -76 \\
\hline
& 2 & +2 & +19 & +70
\end{array}
$$

In general to divide

$$a_n x^n + a_{n-1} x^{n-1} + \cdots + a_1 x + a_0 \text{ by } x - r$$

one writes down

$$\underline{-r}\ \left| \ a_n + a_{n-1} + \cdots + a_1 + a_0 \right.$$

and then proceeds as follows:

Multiply $-r$ times $a_n$ and subtract it from $a_{n-1}$.

$$
\begin{array}{r|l}
-r & a_n + a_{n-1} + \cdots + a_1 + a_0 \\
& \underline{ra_n} \\
& a_n + (a_{n-1} - ra_n)
\end{array}
$$

Then multiply $-r$ times $(a_{n-1} - ra_n)$ and subtract from $a_{n-2}$, etc.

For example, to divide $x^4 - 2x^2 + x + 1$ by $x + 3$, we write

$$
\begin{array}{r|rrrrr}
3 & 1 + 0 - 2 + & 1 + & 1 \\
& \ \ \ \ \ \ 3 - 9 + 21 - & 60 \\
\hline
& 1 - 3 + 7 - 20 + & 61
\end{array}
$$

This says that

$$\frac{x^4 - 2x^2 + x + 1}{x + 3} = x^3 - 3x^2 + 7x - 20 + \frac{61}{x + 3}$$

Note several things:

There was no $x^3$ term in the polynomial, so a 0 coefficient was put in the $x^3$ position.

Since $x - r = x + 3$, the number 3 was placed in the upper left.

The number that appears in the lower right-hand corner is always the remainder in the division.

## ◆ Illustration A.1

Let

$$P(x) = x^3 - 6x^2 - 6$$

Find

$$\frac{P(x)}{x - \frac{5}{2}}$$

$$
\begin{array}{r|rrrr}
-\frac{5}{2} & 1 - 6 + 0 & - 6 \\
& \ \ \ -\frac{5}{2} + \frac{35}{4} + \frac{175}{8} \\
\hline
& 1 - \frac{7}{2} - \frac{35}{4} - \frac{223}{8}
\end{array}
$$

That is,

$$\frac{x^3 - 6x^2 - 6}{x - \frac{5}{2}} = x^2 - \frac{7}{2}x - \frac{35}{4} - \frac{\frac{223}{8}}{x - \frac{5}{2}} \quad \blacktriangleleft$$

A very useful application of synthetic division follows from Theorem 5.2. In short Theorem 5.2 says that if you divide a polynomial $P(x)$ by $x - r$, the remainder is $P(r)$.

Since synthetic division yields the remainder as the lower right-hand number, it can be used as a technique for finding $P(r)$. This is particularly handy if you do not happen to have a hand calculator available.†

♦ **Illustration A.2**

**a** Let $P(x) = 5x^4 - 4x^3 + 2x - 1$. Find $P(0.4)$.

To do this we divide $P(x)$ by $x - 0.4$ and look for the remainder.

$$-0.4 \mid \begin{array}{ccccc} 5 & -4 & +0 & +2 & -1 \\ & -2 & +0.8 & +0.32 & -0.672 \\ \hline 5 & -2 & -0.8 & +1.68 & -0.328 \end{array}$$

So $P(0.4) = -0.328$.

**b** Let $P(x) = 4x^3 - 2x^2 + 3x - 5$. Find $P(-3)$.

To do this we divide by $x - (-3) = x + 3$.

$$3 \mid \begin{array}{cccc} 4 & -2 & +3 & -5 \\ & 12 & -42 & +135 \\ \hline 4 & -14 & +45 & -140 \end{array}$$

So $P(-3) = -140$. ♦

**PROBLEMS**

*Use synthetic division to do each of the following.*

**1** Find $\dfrac{x^4 + 2x^2 - 3x + 4}{x - 3}$  **2** Find $\dfrac{3x^4 + 2x^2 + x}{x + \frac{1}{2}}$

**3** Find $\dfrac{x^4 - 2x^3 + 3x - 1}{x - 4}$  **4** Find $\dfrac{x^5 - 7}{x + \frac{3}{2}}$

**5** Interpret the remainder in each of Probs. 1 to 4 as $P(r)$ for some polynomial $P(x)$ and some number $r$.

**6** Let $P(x) = 3x^4 - 2x^2 + 1$. Find $P(2)$.

**7** Let $P(x) = x^3 + x^2 + x + 1$. Find $P(-0.2)$.

**8** Let $P(x) = -5x^5 + 2x^2 - 6$. Find $P(4)$.

**9** One can devise a procedure for synthetic division in which $r$ and not $-r$ is used and in which terms are added instead of subtracted. See if you can devise such a procedure and apply the procedure to several of the above problems.

† If your hand calculator does not have a memory, you may do well to use it in conjunction with synthetic division.

# B

APPENDIX

# TABLES

| $x$ | $e^x$ | $e^{-x}$ |
|------|---------|----------|
| 0.00 | 1.0000 | 1.00000 |
| 0.01 | 1.0101 | 0.99005 |
| 0.02 | 1.0202 | 0.98020 |
| 0.03 | 1.0305 | 0.97045 |
| 0.04 | 1.0408 | 0.96079 |
| 0.05 | 1.0513 | 0.95123 |
| 0.06 | 1.0618 | 0.94176 |
| 0.07 | 1.0725 | 0.93239 |
| 0.08 | 1.0833 | 0.92312 |
| 0.09 | 1.0942 | 0.91393 |
| 0.10 | 1.1052 | 0.90484 |
| 0.20 | 1.2214 | 0.81873 |
| 0.30 | 1.3499 | 0.74082 |
| 0.40 | 1.4918 | 0.67032 |
| 0.50 | 1.6487 | 0.60653 |
| 0.60 | 1.8221 | 0.54881 |
| 0.70 | 2.0138 | 0.49659 |
| 0.80 | 2.2255 | 0.44933 |
| 0.90 | 2.4596 | 0.40657 |
| 1.00 | 2.7183 | 0.36788 |
| 1.10 | 3.0042 | 0.33287 |
| 1.20 | 3.3201 | 0.30119 |
| 1.30 | 3.6693 | 0.27253 |
| 1.40 | 4.0552 | 0.24660 |
| 1.50 | 4.4817 | 0.22313 |
| 1.60 | 4.9530 | 0.20190 |
| 1.70 | 5.4739 | 0.18268 |
| 1.80 | 6.0496 | 0.16530 |
| 1.90 | 6.6859 | 0.14957 |
| 2.00 | 7.3891 | 0.13534 |
| 2.10 | 8.1662 | 0.12246 |
| 2.20 | 9.0250 | 0.11080 |
| 2.30 | 9.9742 | 0.10026 |
| 2.40 | 11.023 | 0.09072 |
| 2.50 | 12.182 | 0.08208 |
| 2.60 | 13.464 | 0.07427 |
| 2.70 | 14.880 | 0.06721 |
| 2.80 | 16.445 | 0.06081 |
| 2.90 | 18.174 | 0.05502 |
| 3.00 | 20.086 | 0.04979 |
| 3.10 | 22.198 | 0.04505 |
| 3.20 | 24.533 | 0.04076 |
| 3.30 | 27.113 | 0.03688 |
| 3.40 | 29.964 | 0.03337 |
| 3.50 | 33.115 | 0.03020 |
| 3.60 | 36.598 | 0.02732 |
| 3.70 | 40.447 | 0.02472 |
| 3.80 | 44.701 | 0.02237 |
| 3.90 | 49.402 | 0.02024 |
| 4.00 | 54.598 | 0.01832 |
| 4.10 | 60.340 | 0.01657 |
| 4.20 | 66.686 | 0.01500 |
| 4.30 | 73.700 | 0.01357 |
| 4.40 | 81.451 | 0.01228 |
| 4.50 | 90.017 | 0.01111 |
| 4.60 | 99.484 | 0.01005 |
| 4.70 | 109.95 | 0.00910 |
| 4.80 | 121.51 | 0.00823 |
| 4.90 | 134.29 | 0.00745 |
| 5.00 | 148.41 | 0.00674 |

**Table I**
Some
Important
Constants

$$\pi = 3.14159\ 26536$$
$$e = 2.71828\ 18285$$
$$\log_{10} e = 0.43429\ 44819$$
$$\log_e 10 = 2.30258\ 50930$$
$$\pi \text{ radians} = 180°$$
$$1 \text{ radian} = 57.29578°$$
$$= 57°17.74677'$$
$$1° = 0.01745\ 32925 \text{ radian}$$
$$1' = 0.00029\ 08882 \text{ radian}$$

**Table III**
Common
Logarithms
(Base 10)

| N | 0 | 1 | 2 | 3 | 4 | 5 | 6 | 7 | 8 | 9 |
|---|---|---|---|---|---|---|---|---|---|---|
| 10 | 0000 | 0043 | 0086 | 0128 | 0170 | 0212 | 0253 | 0294 | 0334 | 0374 |
| 11 | 0414 | 0453 | 0492 | 0531 | 0569 | 0607 | 0645 | 0682 | 0719 | 0755 |
| 12 | 0792 | 0828 | 0864 | 0899 | 0934 | 0969 | 1004 | 1038 | 1072 | 1106 |
| 13 | 1139 | 1173 | 1206 | 1239 | 1271 | 1303 | 1335 | 1367 | 1399 | 1430 |
| 14 | 1461 | 1492 | 1523 | 1553 | 1584 | 1614 | 1644 | 1673 | 1703 | 1732 |
| 15 | 1761 | 1790 | 1818 | 1847 | 1875 | 1903 | 1931 | 1959 | 1987 | 2014 |
| 16 | 2041 | 2068 | 2095 | 2122 | 2148 | 2175 | 2201 | 2227 | 2253 | 2279 |
| 17 | 2304 | 2330 | 2355 | 2380 | 2405 | 2430 | 2455 | 2480 | 2504 | 2529 |
| 18 | 2553 | 2577 | 2601 | 2625 | 2648 | 2672 | 2695 | 2718 | 2742 | 2765 |
| 19 | 2788 | 2810 | 2833 | 2856 | 2878 | 2900 | 2923 | 2945 | 2967 | 2989 |
| 20 | 3010 | 3032 | 3054 | 3075 | 3096 | 3118 | 3139 | 3160 | 3181 | 3201 |
| 21 | 3222 | 3243 | 3263 | 3284 | 3304 | 3324 | 3345 | 3365 | 3385 | 3404 |
| 22 | 3424 | 3444 | 3464 | 3483 | 3502 | 3522 | 3541 | 3560 | 3579 | 3598 |
| 23 | 3617 | 3636 | 3655 | 3674 | 3692 | 3711 | 3729 | 3747 | 3766 | 3784 |
| 24 | 3802 | 3820 | 3838 | 3856 | 3874 | 3892 | 3909 | 3927 | 3945 | 3962 |
| 25 | 3979 | 3997 | 4014 | 4031 | 4048 | 4065 | 4082 | 4099 | 4116 | 4133 |
| 26 | 4150 | 4166 | 4183 | 4200 | 4216 | 4232 | 4249 | 4265 | 4281 | 4298 |
| 27 | 4314 | 4330 | 4346 | 4362 | 4378 | 4393 | 4409 | 4425 | 4440 | 4456 |
| 28 | 4472 | 4487 | 4502 | 4518 | 4533 | 4548 | 4564 | 4579 | 4594 | 4609 |
| 29 | 4624 | 4639 | 4654 | 4669 | 4683 | 4698 | 4713 | 4728 | 4742 | 4757 |
| 30 | 4771 | 4786 | 4800 | 4814 | 4829 | 4843 | 4857 | 4871 | 4886 | 4900 |
| 31 | 4914 | 4928 | 4942 | 4955 | 4969 | 4983 | 4997 | 5011 | 5024 | 5038 |
| 32 | 5051 | 5065 | 5079 | 5092 | 5105 | 5119 | 5132 | 5145 | 5159 | 5172 |
| 33 | 5185 | 5198 | 5211 | 5224 | 5237 | 5250 | 5263 | 5276 | 5289 | 5302 |
| 34 | 5315 | 5328 | 5340 | 5353 | 5366 | 5378 | 5391 | 5403 | 5416 | 5428 |
| 35 | 5441 | 5453 | 5465 | 5478 | 5490 | 5502 | 5514 | 5527 | 5539 | 5551 |
| 36 | 5563 | 5575 | 5587 | 5599 | 5611 | 5623 | 5635 | 5647 | 5658 | 5670 |
| 37 | 5682 | 5694 | 5705 | 5717 | 5729 | 5740 | 5752 | 5763 | 5775 | 5786 |
| 38 | 5798 | 5809 | 5821 | 5832 | 5843 | 5855 | 5866 | 5877 | 5888 | 5899 |
| 39 | 5911 | 5922 | 5933 | 5944 | 5955 | 5966 | 5977 | 5988 | 5999 | 6010 |
| 40 | 6021 | 6031 | 6042 | 6053 | 6064 | 6075 | 6085 | 6096 | 6107 | 6117 |
| 41 | 6128 | 6138 | 6149 | 6160 | 6170 | 6180 | 6191 | 6201 | 6212 | 6222 |
| 42 | 6232 | 6243 | 6253 | 6263 | 6274 | 6284 | 6294 | 6304 | 6314 | 6325 |
| 43 | 6335 | 6345 | 6355 | 6365 | 6375 | 6385 | 6395 | 6405 | 6415 | 6425 |
| 44 | 6435 | 6444 | 6454 | 6464 | 6474 | 6484 | 6493 | 6503 | 6513 | 6522 |
| 45 | 6532 | 6542 | 6551 | 6561 | 6571 | 6580 | 6590 | 6599 | 6609 | 6618 |
| 46 | 6628 | 6637 | 6646 | 6656 | 6665 | 6675 | 6684 | 6693 | 6702 | 6712 |
| 47 | 6721 | 6730 | 6739 | 6749 | 6758 | 6767 | 6776 | 6785 | 6794 | 6803 |
| 48 | 6812 | 6821 | 6830 | 6839 | 6848 | 6857 | 6866 | 6875 | 6884 | 6893 |
| 49 | 6902 | 6911 | 6920 | 6928 | 6937 | 6946 | 6955 | 6964 | 6972 | 6981 |
| 50 | 6990 | 6998 | 7007 | 7016 | 7024 | 7033 | 7042 | 7050 | 7059 | 7067 |
| 51 | 7076 | 7084 | 7093 | 7101 | 7110 | 7118 | 7126 | 7135 | 7143 | 7152 |
| 52 | 7160 | 7168 | 7177 | 7185 | 7193 | 7202 | 7210 | 7218 | 7226 | 7235 |
| 53 | 7243 | 7251 | 7259 | 7267 | 7275 | 7284 | 7292 | 7300 | 7308 | 7316 |
| 54 | 7324 | 7332 | 7340 | 7348 | 7356 | 7364 | 7372 | 7380 | 7388 | 7396 |
| N | 0 | 1 | 2 | 3 | 4 | 5 | 6 | 7 | 8 | 9 |

**Table III**
Common
Logarithms
(continued)

| N | 0 | 1 | 2 | 3 | 4 | 5 | 6 | 7 | 8 | 9 |
|---|---|---|---|---|---|---|---|---|---|---|
| 55 | 7404 | 7412 | 7419 | 7427 | 7435 | 7443 | 7451 | 7459 | 7466 | 7474 |
| 56 | 7482 | 7490 | 7497 | 7505 | 7513 | 7520 | 7528 | 7536 | 7543 | 7551 |
| 57 | 7559 | 7566 | 7574 | 7582 | 7589 | 7597 | 7604 | 7612 | 7619 | 7627 |
| 58 | 7634 | 7642 | 7649 | 7657 | 7664 | 7672 | 7679 | 7686 | 7694 | 7701 |
| 59 | 7709 | 7716 | 7723 | 7731 | 7738 | 7745 | 7752 | 7760 | 7767 | 7774 |
| 60 | 7782 | 7789 | 7796 | 7803 | 7810 | 7818 | 7825 | 7832 | 7839 | 7846 |
| 61 | 7853 | 7860 | 7868 | 7875 | 7882 | 7889 | 7896 | 7903 | 7910 | 7917 |
| 62 | 7924 | 7931 | 7938 | 7945 | 7952 | 7959 | 7966 | 7973 | 7980 | 7987 |
| 63 | 7993 | 8000 | 8007 | 8014 | 8021 | 8028 | 8035 | 8041 | 8048 | 8055 |
| 64 | 8062 | 8069 | 8075 | 8082 | 8089 | 8096 | 8102 | 8109 | 8116 | 8122 |
| 65 | 8129 | 8136 | 8142 | 8149 | 8156 | 8162 | 8169 | 8176 | 8182 | 8189 |
| 66 | 8195 | 8202 | 8209 | 8215 | 8222 | 8228 | 8235 | 8241 | 8248 | 8254 |
| 67 | 8261 | 8267 | 8274 | 8280 | 8287 | 8293 | 8299 | 8306 | 8312 | 8319 |
| 68 | 8325 | 8331 | 8338 | 8344 | 8351 | 8357 | 8363 | 8370 | 8376 | 8382 |
| 69 | 8388 | 8395 | 8401 | 8407 | 8414 | 8420 | 8426 | 8432 | 8439 | 8445 |
| 70 | 8451 | 8457 | 8463 | 8470 | 8476 | 8482 | 8488 | 8494 | 8500 | 8506 |
| 71 | 8513 | 8519 | 8525 | 8531 | 8537 | 8543 | 8549 | 8555 | 8561 | 8567 |
| 72 | 8573 | 8579 | 8585 | 8591 | 8597 | 8603 | 8609 | 8615 | 8621 | 8627 |
| 73 | 8633 | 8639 | 8645 | 8651 | 8657 | 8663 | 8669 | 8675 | 8681 | 8686 |
| 74 | 8692 | 8698 | 8704 | 8710 | 8716 | 8722 | 8727 | 8733 | 8739 | 8745 |
| 75 | 8751 | 8756 | 8762 | 8768 | 8774 | 8779 | 8785 | 8791 | 8797 | 8802 |
| 76 | 8808 | 8814 | 8820 | 8825 | 8831 | 8837 | 8842 | 8848 | 8854 | 8859 |
| 77 | 8865 | 8871 | 8876 | 8882 | 8887 | 8893 | 8899 | 8904 | 8910 | 8915 |
| 78 | 8921 | 8927 | 8932 | 8938 | 8943 | 8949 | 8954 | 8960 | 8965 | 8971 |
| 79 | 8976 | 8982 | 8987 | 8993 | 8998 | 9004 | 9009 | 9015 | 9020 | 9025 |
| 80 | 9031 | 9036 | 9042 | 9047 | 9053 | 9058 | 9063 | 9069 | 9074 | 9079 |
| 81 | 9085 | 9090 | 9096 | 9101 | 9106 | 9112 | 9117 | 9122 | 9128 | 9133 |
| 82 | 9138 | 9143 | 9149 | 9154 | 9159 | 9165 | 9170 | 9175 | 9180 | 9186 |
| 83 | 9191 | 9196 | 9201 | 9206 | 9212 | 9217 | 9222 | 9227 | 9232 | 9238 |
| 84 | 9243 | 9248 | 9253 | 9258 | 9263 | 9269 | 9274 | 9279 | 9284 | 9289 |
| 85 | 9294 | 9299 | 9304 | 9309 | 9315 | 9320 | 9325 | 9330 | 9335 | 9340 |
| 86 | 9345 | 9350 | 9355 | 9360 | 9365 | 9370 | 9375 | 9380 | 9385 | 9390 |
| 87 | 9395 | 9400 | 9405 | 9410 | 9415 | 9420 | 9425 | 9430 | 9435 | 9440 |
| 88 | 9445 | 9450 | 9455 | 9460 | 9465 | 9469 | 9474 | 9479 | 9484 | 9489 |
| 89 | 9494 | 9499 | 9504 | 9509 | 9513 | 9518 | 9523 | 9528 | 9533 | 9538 |
| 90 | 9542 | 9547 | 9552 | 9557 | 9562 | 9566 | 9571 | 9576 | 9581 | 9586 |
| 91 | 9590 | 9595 | 9600 | 9605 | 9609 | 9614 | 9619 | 9624 | 9628 | 9633 |
| 92 | 9638 | 9643 | 9647 | 9652 | 9657 | 9661 | 9666 | 9671 | 9675 | 9680 |
| 93 | 9685 | 9689 | 9694 | 9699 | 9703 | 9708 | 9713 | 9717 | 9722 | 9727 |
| 94 | 9731 | 9736 | 9741 | 9745 | 9750 | 9754 | 9759 | 9763 | 9768 | 9773 |
| 95 | 9777 | 9782 | 9786 | 9791 | 9795 | 9800 | 9805 | 9809 | 9814 | 9818 |
| 96 | 9823 | 9827 | 9832 | 9836 | 9841 | 9845 | 9850 | 9854 | 9859 | 9863 |
| 97 | 9868 | 9872 | 9877 | 9881 | 9886 | 9890 | 9894 | 9899 | 9903 | 9908 |
| 98 | 9912 | 9917 | 9921 | 9926 | 9930 | 9934 | 9939 | 9943 | 9948 | 9952 |
| 99 | 9956 | 9961 | 9965 | 9969 | 9974 | 9978 | 9983 | 9987 | 9991 | 9996 |
| N | 0 | 1 | 2 | 3 | 4 | 5 | 6 | 7 | 8 | 9 |

**Table IV**
Natural
Logarithms
(Base e)

| N | .00 | .01 | .02 | .03 | .04 | .05 | .06 | .07 | .08 | .09 |
|---|-----|-----|-----|-----|-----|-----|-----|-----|-----|-----|
| 1.0 | 0.0000 | 0.0100 | 0.0198 | 0.0296 | 0.0392 | 0.0488 | 0.0583 | 0.0677 | 0.0770 | 0.0862 |
| 1.1 | 0.0953 | 0.1044 | 0.1133 | 0.1222 | 0.1310 | 0.1398 | 0.1484 | 0.1570 | 0.1655 | 0.1740 |
| 1.2 | 0.1823 | 0.1906 | 0.1989 | 0.2070 | 0.2151 | 0.2231 | 0.2311 | 0.2390 | 0.2469 | 0.2546 |
| 1.3 | 0.2624 | 0.2700 | 0.2776 | 0.2852 | 0.2927 | 0.3001 | 0.3075 | 0.3148 | 0.3221 | 0.3293 |
| 1.4 | 0.3365 | 0.3436 | 0.3507 | 0.3577 | 0.3646 | 0.3716 | 0.3784 | 0.3853 | 0.3920 | 0.3988 |
| 1.5 | 0.4055 | 0.4121 | 0.4187 | 0.4253 | 0.4318 | 0.4383 | 0.4447 | 0.4511 | 0.4574 | 0.4637 |
| 1.6 | 0.4700 | 0.4762 | 0.4824 | 0.4886 | 0.4947 | 0.5008 | 0.5068 | 0.5128 | 0.5188 | 0.5247 |
| 1.7 | 0.5306 | 0.5365 | 0.5423 | 0.5481 | 0.5539 | 0.5596 | 0.5653 | 0.5710 | 0.5766 | 0.5822 |
| 1.8 | 0.5878 | 0.5933 | 0.5988 | 0.6043 | 0.6098 | 0.6152 | 0.6206 | 0.6259 | 0.6313 | 0.6366 |
| 1.9 | 0.6419 | 0.6471 | 0.6523 | 0.6575 | 0.6627 | 0.6678 | 0.6729 | 0.6780 | 0.6831 | 0.6881 |
| 2.0 | 0.6932 | 0.6981 | 0.7031 | 0.7080 | 0.7129 | 0.7178 | 0.7227 | 0.7275 | 0.7324 | 0.7372 |
| 2.1 | 0.7419 | 0.7467 | 0.7514 | 0.7561 | 0.7608 | 0.7655 | 0.7701 | 0.7747 | 0.7793 | 0.7839 |
| 2.2 | 0.7885 | 0.7930 | 0.7975 | 0.8020 | 0.8065 | 0.8109 | 0.8154 | 0.8198 | 0.8242 | 0.8286 |
| 2.3 | 0.8329 | 0.8373 | 0.8416 | 0.8459 | 0.8502 | 0.8544 | 0.8587 | 0.8629 | 0.8671 | 0.8713 |
| 2.4 | 0.8755 | 0.8796 | 0.8838 | 0.8879 | 0.8920 | 0.8961 | 0.9002 | 0.9042 | 0.9083 | 0.9123 |
| 2.5 | 0.9163 | 0.9203 | 0.9243 | 0.9282 | 0.9322 | 0.9361 | 0.9400 | 0.9439 | 0.9478 | 0.9517 |
| 2.6 | 0.9555 | 0.9594 | 0.9632 | 0.9670 | 0.9708 | 0.9746 | 0.9783 | 0.9821 | 0.9858 | 0.9895 |
| 2.7 | 0.9933 | 0.9969 | 1.0006 | 1.0043 | 1.0080 | 1.0116 | 1.0152 | 1.0188 | 1.0225 | 1.0260 |
| 2.8 | 1.0296 | 1.0332 | 1.0367 | 1.0403 | 1.0438 | 1.0473 | 1.0508 | 1.0543 | 1.0578 | 1.0613 |
| 2.9 | 1.0647 | 1.0682 | 1.0716 | 1.0750 | 1.0784 | 1.0818 | 1.0852 | 1.0886 | 1.0919 | 1.0953 |
| 3.0 | 1.0986 | 1.1019 | 1.1053 | 1.1086 | 1.1119 | 1.1151 | 1.1184 | 1.1217 | 1.1249 | 1.1282 |
| 3.1 | 1.1314 | 1.1346 | 1.1378 | 1.1410 | 1.1442 | 1.1474 | 1.1506 | 1.1537 | 1.1569 | 1.1600 |
| 3.2 | 1.1632 | 1.1663 | 1.1694 | 1.1725 | 1.1756 | 1.1787 | 1.1817 | 1.1848 | 1.1878 | 1.1909 |
| 3.3 | 1.1939 | 1.1969 | 1.2000 | 1.2030 | 1.2060 | 1.2090 | 1.2119 | 1.2149 | 1.2179 | 1.2208 |
| 3.4 | 1.2238 | 1.2267 | 1.2296 | 1.2326 | 1.2355 | 1.2384 | 1.2413 | 1.2442 | 1.2470 | 1.2499 |
| 3.5 | 1.2528 | 1.2556 | 1.2585 | 1.2613 | 1.2641 | 1.2669 | 1.2698 | 1.2726 | 1.2754 | 1.2782 |
| 3.6 | 1.2809 | 1.2837 | 1.2865 | 1.2892 | 1.2920 | 1.2947 | 1.2975 | 1.3002 | 1.3029 | 1.3056 |
| 3.7 | 1.3083 | 1.3110 | 1.3137 | 1.3164 | 1.3191 | 1.3218 | 1.3244 | 1.3271 | 1.3297 | 1.3324 |
| 3.8 | 1.3350 | 1.3376 | 1.3403 | 1.3429 | 1.3455 | 1.3481 | 1.3507 | 1.3533 | 1.3558 | 1.3584 |
| 3.9 | 1.3610 | 1.3635 | 1.3661 | 1.3686 | 1.3712 | 1.3737 | 1.3762 | 1.3788 | 1.3813 | 1.3838 |
| 4.0 | 1.3863 | 1.3888 | 1.3913 | 1.3938 | 1.3962 | 1.3987 | 1.4012 | 1.4036 | 1.4061 | 1.4085 |
| 4.1 | 1.4110 | 1.4134 | 1.4159 | 1.4183 | 1.4207 | 1.4231 | 1.4255 | 1.4279 | 1.4303 | 1.4327 |
| 4.2 | 1.4351 | 1.4375 | 1.4398 | 1.4422 | 1.4446 | 1.4469 | 1.4493 | 1.4516 | 1.4540 | 1.4563 |
| 4.3 | 1.4586 | 1.4609 | 1.4633 | 1.4656 | 1.4679 | 1.4702 | 1.4725 | 1.4748 | 1.4771 | 1.4793 |
| 4.4 | 1.4816 | 1.4839 | 1.4861 | 1.4884 | 1.4907 | 1.4929 | 1.4951 | 1.4974 | 1.4996 | 1.5019 |
| 4.5 | 1.5041 | 1.5063 | 1.5085 | 1.5107 | 1.5129 | 1.5151 | 1.5173 | 1.5195 | 1.5217 | 1.5239 |
| 4.6 | 1.5261 | 1.5282 | 1.5304 | 1.5326 | 1.5347 | 1.5369 | 1.5390 | 1.5412 | 1.5433 | 1.5454 |
| 4.7 | 1.5476 | 1.5497 | 1.5518 | 1.5539 | 1.5560 | 1.5581 | 1.5602 | 1.5623 | 1.5644 | 1.5665 |
| 4.8 | 1.5686 | 1.5707 | 1.5728 | 1.5748 | 1.5769 | 1.5790 | 1.5810 | 1.5831 | 1.5851 | 1.5872 |
| 4.9 | 1.5892 | 1.5913 | 1.5933 | 1.5953 | 1.5974 | 1.5994 | 1.6014 | 1.6034 | 1.6054 | 1.6074 |
| 5.0 | 1.6094 | 1.6114 | 1.6134 | 1.6154 | 1.6174 | 1.6194 | 1.6214 | 1.6233 | 1.6253 | 1.6273 |
| 5.1 | 1.6292 | 1.6312 | 1.6332 | 1.6351 | 1.6371 | 1.6390 | 1.6409 | 1.6429 | 1.6448 | 1.6467 |
| 5.2 | 1.6487 | 1.6506 | 1.6525 | 1.6544 | 1.6563 | 1.6582 | 1.6601 | 1.6620 | 1.6639 | 1.6658 |
| 5.3 | 1.6677 | 1.6696 | 1.6715 | 1.6734 | 1.6752 | 1.6771 | 1.6790 | 1.6808 | 1.6827 | 1.6845 |
| 5.4 | 1.6864 | 1.6882 | 1.6901 | 1.6919 | 1.6938 | 1.6956 | 1.6974 | 1.6993 | 1.7011 | 1.7029 |
| N | .00 | .01 | .02 | .03 | .04 | .05 | .06 | .07 | .08 | .09 |

**Table IV**
Natural
Logarithms
(continued)

| N | .00 | .01 | .02 | .03 | .04 | .05 | .06 | .07 | .08 | .09 |
|---|------|------|------|------|------|------|------|------|------|------|
| 5.5 | 1.7047 | 1.7066 | 1.7084 | 1.7102 | 1.7120 | 1.7138 | 1.7156 | 1.7174 | 1.7192 | 1.7210 |
| 5.6 | 1.7228 | 1.7246 | 1.7263 | 1.7281 | 1.7299 | 1.7317 | 1.7334 | 1.7352 | 1.7370 | 1.7387 |
| 5.7 | 1.7405 | 1.7422 | 1.7440 | 1.7457 | 1.7475 | 1.7492 | 1.7509 | 1.7527 | 1.7544 | 1.7561 |
| 5.8 | 1.7579 | 1.7596 | 1.7613 | 1.7630 | 1.7647 | 1.7664 | 1.7681 | 1.7699 | 1.7716 | 1.7733 |
| 5.9 | 1.7750 | 1.7766 | 1.7783 | 1.7800 | 1.7817 | 1.7834 | 1.7851 | 1.7868 | 1.7884 | 1.7901 |
| 6.0 | 1.7918 | 1.7934 | 1.7951 | 1.7967 | 1.7984 | 1.8001 | 1.8017 | 1.8034 | 1.8050 | 1.8066 |
| 6.1 | 1.8083 | 1.8099 | 1.8116 | 1.8132 | 1.8148 | 1.8165 | 1.8181 | 1.8197 | 1.8213 | 1.8229 |
| 6.2 | 1.8245 | 1.8262 | 1.8278 | 1.8294 | 1.8310 | 1.8326 | 1.8342 | 1.8358 | 1.8374 | 1.8390 |
| 6.3 | 1.8405 | 1.8421 | 1.8437 | 1.8453 | 1.8469 | 1.8485 | 1.8500 | 1.8516 | 1.8532 | 1.8547 |
| 6.4 | 1.8563 | 1.8579 | 1.8594 | 1.8610 | 1.8625 | 1.8641 | 1.8656 | 1.8672 | 1.8687 | 1.8703 |
| 6.5 | 1.8718 | 1.8733 | 1.8749 | 1.8764 | 1.8779 | 1.8795 | 1.8810 | 1.8825 | 1.8840 | 1.8856 |
| 6.6 | 1.8871 | 1.8886 | 1.8901 | 1.8916 | 1.8931 | 1.8946 | 1.8961 | 1.8976 | 1.8991 | 1.9006 |
| 6.7 | 1.9021 | 1.9036 | 1.9051 | 1.9066 | 1.9081 | 1.9095 | 1.9110 | 1.9125 | 1.9140 | 1.9155 |
| 6.8 | 1.9169 | 1.9184 | 1.9199 | 1.9213 | 1.9228 | 1.9242 | 1.9257 | 1.9272 | 1.9286 | 1.9301 |
| 6.9 | 1.9315 | 1.9330 | 1.9344 | 1.9359 | 1.9373 | 1.9387 | 1.9402 | 1.9416 | 1.9430 | 1.9445 |
| 7.0 | 1.9459 | 1.9473 | 1.9488 | 1.9502 | 1.9516 | 1.9530 | 1.9544 | 1.9559 | 1.9573 | 1.9587 |
| 7.1 | 1.9601 | 1.9615 | 1.9629 | 1.9643 | 1.9657 | 1.9671 | 1.9685 | 1.9699 | 1.9713 | 1.9727 |
| 7.2 | 1.9741 | 1.9755 | 1.9769 | 1.9782 | 1.9796 | 1.9810 | 1.9824 | 1.9838 | 1.9851 | 1.9865 |
| 7.3 | 1.9879 | 1.9892 | 1.9906 | 1.9920 | 1.9933 | 1.9947 | 1.9961 | 1.9974 | 1.9988 | 2.0001 |
| 7.4 | 2.0015 | 2.0028 | 2.0042 | 2.0055 | 2.0069 | 2.0082 | 2.0096 | 2.0109 | 2.0122 | 2.0136 |
| 7.5 | 2.0149 | 2.0162 | 2.0176 | 2.0189 | 2.0202 | 2.0215 | 2.0229 | 2.0242 | 2.0255 | 2.0268 |
| 7.6 | 2.0281 | 2.0295 | 2.0308 | 2.0321 | 2.0334 | 2.0347 | 2.0360 | 2.0373 | 2.0386 | 2.0399 |
| 7.7 | 2.0412 | 2.0425 | 2.0438 | 2.0451 | 2.0464 | 2.0477 | 2.0490 | 2.0503 | 2.0516 | 2.0528 |
| 7.8 | 2.0541 | 2.0554 | 2.0567 | 2.0580 | 2.0592 | 2.0605 | 2.0618 | 2.0631 | 2.0643 | 2.0656 |
| 7.9 | 2.0669 | 2.0681 | 2.0694 | 2.0707 | 2.0719 | 2.0732 | 2.0744 | 2.0757 | 2.0769 | 2.0782 |
| 8.0 | 2.0794 | 2.0807 | 2.0819 | 2.0832 | 2.0844 | 2.0857 | 2.0869 | 2.0882 | 2.0894 | 2.0906 |
| 8.1 | 2.0919 | 2.0931 | 2.0943 | 2.0956 | 2.0968 | 2.0980 | 2.0992 | 2.1005 | 2.1017 | 2.1029 |
| 8.2 | 2.1041 | 2.1054 | 2.1066 | 2.1078 | 2.1090 | 2.1102 | 2.1114 | 2.1126 | 2.1138 | 2.1150 |
| 8.3 | 2.1163 | 2.1175 | 2.1187 | 2.1199 | 2.1211 | 2.1223 | 2.1235 | 2.1247 | 2.1259 | 2.1270 |
| 8.4 | 2.1282 | 2.1294 | 2.1306 | 2.1318 | 2.1330 | 2.1342 | 2.1353 | 2.1365 | 2.1377 | 2.1389 |
| 8.5 | 2.1401 | 2.1412 | 2.1424 | 2.1436 | 2.1448 | 2.1459 | 2.1471 | 2.1483 | 2.1494 | 2.1506 |
| 8.6 | 2.1518 | 2.1529 | 2.1541 | 2.1552 | 2.1564 | 2.1576 | 2.1587 | 2.1599 | 2.1610 | 2.1622 |
| 8.7 | 2.1633 | 2.1645 | 2.1656 | 2.1668 | 2.1679 | 2.1691 | 2.1702 | 2.1713 | 2.1725 | 2.1736 |
| 8.8 | 2.1748 | 2.1759 | 2.1770 | 2.1782 | 2.1793 | 2.1804 | 2.1815 | 2.1827 | 2.1838 | 2.1849 |
| 8.9 | 2.1861 | 2.1872 | 2.1883 | 2.1894 | 2.1905 | 2.1917 | 2.1928 | 2.1939 | 2.1950 | 2.1961 |
| 9.0 | 2.1972 | 2.1983 | 2.1994 | 2.2006 | 2.2017 | 2.2028 | 2.2039 | 2.2050 | 2.2061 | 2.2072 |
| 9.1 | 2.2083 | 2.2094 | 2.2105 | 2.2116 | 2.2127 | 2.2138 | 2.2148 | 2.2159 | 2.2170 | 2.2181 |
| 9.2 | 2.2192 | 2.2203 | 2.2214 | 2.2225 | 2.2235 | 2.2246 | 2.2257 | 2.2268 | 2.2279 | 2.2289 |
| 9.3 | 2.2300 | 2.2311 | 2.2322 | 2.2332 | 2.2343 | 2.2354 | 2.2364 | 2.2375 | 2.2386 | 2.2396 |
| 9.4 | 2.2407 | 2.2418 | 2.2428 | 2.2439 | 2.2450 | 2.2460 | 2.2471 | 2.2481 | 2.2492 | 2.2502 |
| 9.5 | 2.2513 | 2.2523 | 2.2534 | 2.2544 | 2.2555 | 2.2565 | 2.2576 | 2.2586 | 2.2597 | 2.2607 |
| 9.6 | 2.2618 | 2.2628 | 2.2638 | 2.2649 | 2.2659 | 2.2670 | 2.2680 | 2.2690 | 2.2701 | 2.2711 |
| 9.7 | 2.2721 | 2.2732 | 2.2742 | 2.2752 | 2.2762 | 2.2773 | 2.2783 | 2.2793 | 2.2803 | 2.2814 |
| 9.8 | 2.2824 | 2.2834 | 2.2844 | 2.2854 | 2.2865 | 2.2875 | 2.2885 | 2.2895 | 2.2905 | 2.2915 |
| 9.9 | 2.2925 | 2.2935 | 2.2946 | 2.2956 | 2.2966 | 2.2976 | 2.2986 | 2.2996 | 2.3006 | 2.3016 |
| N | .00 | .01 | .02 | .03 | .04 | .05 | .06 | .07 | .08 | .09 |

**Table IV**
Natural
Logarithms
(continued)

| N | Nat Log | N | Nat Log | N | Nat Log | N | Nat Log | N | Nat Log |
|---|---------|---|---------|---|---------|---|---------|---|---------|
| 0 | $-\infty$ | 40 | 3.68 888 | 80 | 4.38 203 | 120 | 4.78 749 | 160 | 5.07 517 |
| 1 | 0.00 000 | 41 | 3.71 357 | 81 | 4.39 445 | 121 | 4.79 579 | 161 | 5.08 140 |
| 2 | 0.69 315 | 42 | 3.73 767 | 82 | 4.40 672 | 122 | 4.80 402 | 162 | 5.08 760 |
| 3 | 1.09 861 | 43 | 3.76 120 | 83 | 4.41 884 | 123 | 4.81 218 | 163 | 5.09 375 |
| 4 | 1.38 629 | 44 | 3.78 419 | 84 | 4.43 082 | 124 | 4.82 028 | 164 | 5.09 987 |
| 5 | 1.60 944 | 45 | 3.80 666 | 85 | 4.44 265 | 125 | 4.82 831 | 165 | 5.10 595 |
| 6 | 1.79 176 | 46 | 3.82 864 | 86 | 4.45 435 | 126 | 4.83 628 | 166 | 5.11 199 |
| 7 | 1.94 591 | 47 | 3.85 015 | 87 | 4.46 591 | 127 | 4.84 419 | 167 | 5.11 799 |
| 8 | 2.07 944 | 48 | 3.87 120 | 88 | 4.47 734 | 128 | 4.85 203 | 168 | 5.12 396 |
| 9 | 2.19 722 | 49 | 3.89 182 | 89 | 4.48 864 | 129 | 4.85 981 | 169 | 5.12 990 |
| 10 | 2.30 259 | 50 | 3.91 202 | 90 | 4.49 981 | 130 | 4.86 753 | 170 | 5.13 580 |
| 11 | 2.39 790 | 51 | 3.93 183 | 91 | 4.51 086 | 131 | 4.87 520 | 171 | 5.14 166 |
| 12 | 2.48 491 | 52 | 3.95 124 | 92 | 4.52 179 | 132 | 4.88 280 | 172 | 5.14 749 |
| 13 | 2.56 495 | 53 | 3.97 029 | 93 | 4.53 260 | 133 | 4.89 035 | 173 | 5.15 329 |
| 14 | 2.63 906 | 54 | 3.98 898 | 94 | 4.54 329 | 134 | 4.89 784 | 174 | 5.15 906 |
| 15 | 2.70 805 | 55 | 4.00 733 | 95 | 4.55 388 | 135 | 4.90 527 | 175 | 5.16 479 |
| 16 | 2.77 259 | 56 | 4.02 535 | 96 | 4.56 435 | 136 | 4.91 265 | 176 | 5.17 048 |
| 17 | 2.83 321 | 57 | 4.04 305 | 97 | 4.57 471 | 137 | 4.91 998 | 177 | 5.17 615 |
| 18 | 2.89 037 | 58 | 4.06 044 | 98 | 4.58 497 | 138 | 4.92 725 | 178 | 5.18 178 |
| 19 | 2.94 444 | 59 | 4.07 754 | 99 | 4.59 512 | 139 | 4.93 447 | 179 | 5.18 739 |
| 20 | 2.99 573 | 60 | 4.09 434 | 100 | 4.60 517 | 140 | 4.94 164 | 180 | 5.19 296 |
| 21 | 3.04 452 | 61 | 4.11 087 | 101 | 4.61 512 | 141 | 4.94 876 | 181 | 5.19 850 |
| 22 | 3.09 104 | 62 | 4.12 713 | 102 | 4.62 497 | 142 | 4.95 583 | 182 | 5.20 401 |
| 23 | 3.13 549 | 63 | 4.14 313 | 103 | 4.63 473 | 143 | 4.96 284 | 183 | 5.20 949 |
| 24 | 3.17 805 | 64 | 4.15 888 | 104 | 4.64 439 | 144 | 4.96 981 | 184 | 5.21 494 |
| 25 | 3.21 888 | 65 | 4.17 439 | 105 | 4.65 396 | 145 | 4.97 673 | 185 | 5.22 036 |
| 26 | 3.25 810 | 66 | 4.18 965 | 106 | 4.66 344 | 146 | 4.98 361 | 186 | 5.22 575 |
| 27 | 3.29 584 | 67 | 4.20 469 | 107 | 4.67 283 | 147 | 4.99 043 | 187 | 5.23 111 |
| 28 | 3.33 220 | 68 | 4.21 951 | 108 | 4.68 213 | 148 | 4.99 721 | 188 | 5.23 644 |
| 29 | 3.36 730 | 69 | 4.23 411 | 109 | 4.69 135 | 149 | 5.00 395 | 189 | 5.24 175 |
| 30 | 3.40 120 | 70 | 4.24 850 | 110 | 4.70 048 | 150 | 5.01 064 | 190 | 5.24 702 |
| 31 | 3.43 399 | 71 | 4.26 268 | 111 | 4.70 953 | 151 | 5.01 728 | 191 | 5.25 227 |
| 32 | 3.46 574 | 72 | 4.27 667 | 112 | 4.71 850 | 152 | 5.02 388 | 192 | 5.25 750 |
| 33 | 3.49 651 | 73 | 4.29 046 | 113 | 4.72 739 | 153 | 5.03 044 | 193 | 5.26 269 |
| 34 | 3.52 636 | 74 | 4.30 407 | 114 | 4.73 620 | 154 | 5.03 695 | 194 | 5.26 786 |
| 35 | 3.55 535 | 75 | 4.31 749 | 115 | 4.74 493 | 155 | 5.04 343 | 195 | 5.27 300 |
| 36 | 3.58 352 | 76 | 4.33 073 | 116 | 4.75 359 | 156 | 5.04 986 | 196 | 5.27 811 |
| 37 | 3.61 092 | 77 | 4.34 381 | 117 | 4.76 217 | 157 | 5.05 625 | 197 | 5.28 320 |
| 38 | 3.63 759 | 78 | 4.35 671 | 118 | 4.77 068 | 158 | 5.06 260 | 198 | 5.28 827 |
| 39 | 3.66 356 | 79 | 4.36 945 | 119 | 4.77 912 | 159 | 5.06 890 | 199 | 5.29 330 |
| 40 | 3.68 888 | 80 | 4.38 203 | 120 | 4.78 749 | 160 | 5.07 517 | 200 | 5.29 832 |
| N | Nat Log | N | Nat Log | N | Nat Log | N | Nat Log | N | Nat Log |

**Table V**
Trigonometric
Functions
of Real
Numbers

| $x$ | Sin $x$ | Tan $x$ | Cot $x$ | Cos $x$ | $x$ | Sin $x$ | Tan $x$ | Cot $x$ | Cos $x$ |
|---|---|---|---|---|---|---|---|---|---|
| **.00** | .00000 | .00000 | $\infty$ | 1.00000 | **.50** | .47943 | .54630 | 1.8305 | .87758 |
| .01 | .01000 | .01000 | 99.997 | .99995 | .51 | .48818 | .55936 | 1.7878 | .87274 |
| .02 | .02000 | .02000 | 49.993 | .99980 | .52 | .49688 | .57256 | 1.7465 | .86782 |
| .03 | .03000 | .03001 | 33.323 | .99955 | .53 | .50553 | .58592 | 1.7067 | .86281 |
| .04 | .03999 | .04002 | 24.987 | .99920 | .54 | .51414 | .59943 | 1.6683 | .85771 |
| .05 | .04998 | .05004 | 19.983 | .99875 | .55 | .52269 | .61311 | 1.6310 | .85252 |
| .06 | .05996 | .06007 | 16.647 | .99820 | .56 | .53119 | .62695 | 1.5950 | .84726 |
| .07 | .06994 | .07011 | 14.262 | .99755 | .57 | .53963 | .64097 | 1.5601 | .84190 |
| .08 | .07991 | .08017 | 12.473 | .99680 | .58 | .54802 | .65517 | 1.5263 | .83646 |
| .09 | .08988 | .09024 | 11.081 | .99595 | .59 | .55636 | .66956 | 1.4935 | .83094 |
| **.10** | .09983 | .10033 | 9.9666 | .99500 | **.60** | .56464 | .68414 | 1.4617 | .82534 |
| .11 | .10978 | .11045 | 9.0542 | .99396 | .61 | .57287 | .69892 | 1.4308 | .81965 |
| .12 | .11971 | .12058 | 8.2933 | .99281 | .62 | .58104 | .71391 | 1.4007 | .81388 |
| .13 | .12963 | .13074 | 7.6489 | .99156 | .63 | .58914 | .72911 | 1.3715 | .80803 |
| .14 | .13954 | .14092 | 7.0961 | .99022 | .64 | .59720 | .74454 | 1.3431 | .80210 |
| .15 | .14944 | .15144 | 6.6166 | .98877 | .65 | .60519 | .76020 | 1.3154 | .79608 |
| .16 | .15932 | .16138 | 6.1966 | .98723 | .66 | .61312 | .77610 | 1.2885 | .78999 |
| .17 | .16918 | .17166 | 5.8256 | .98558 | .67 | .62099 | .79225 | 1.2622 | .78382 |
| .18 | .17903 | .18197 | 5.4954 | .98384 | .68 | .62879 | .80866 | 1.2366 | .77757 |
| .19 | .18886 | .19232 | 5.1997 | .98200 | .69 | .63654 | .82534 | 1.2116 | .77125 |
| **.20** | .19867 | .20271 | 4.9332 | .98007 | **.70** | .64422 | .84229 | 1.1872 | .76484 |
| .21 | .20846 | .21314 | 4.6917 | .97803 | .71 | .65183 | .85953 | 1.1634 | .75836 |
| .22 | .21823 | .22362 | 4.4719 | .97590 | .72 | .65938 | .87707 | 1.1402 | .75181 |
| .23 | .22798 | .23414 | 4.2709 | .97367 | .73 | .66687 | .89492 | 1.1174 | .74517 |
| .24 | .23770 | .24472 | 4.0864 | .97134 | .74 | .67429 | .91309 | 1.0952 | .73847 |
| .25 | .24740 | .25534 | 3.9163 | .96891 | .75 | .68164 | .93160 | 1.0734 | .73169 |
| .26 | .25708 | .26602 | 3.7591 | .96639 | .76 | .68892 | .95045 | 1.0521 | .72484 |
| .27 | .26673 | .27676 | 3.6133 | .96377 | .77 | .69614 | .96967 | 1.0313 | .71791 |
| .28 | .27636 | .28755 | 3.4776 | .96106 | .78 | .70328 | .98926 | 1.0109 | .71091 |
| .29 | .28595 | .29841 | 3.3511 | .95824 | .79 | .71035 | 1.0092 | .99084 | .70385 |
| **.30** | .29552 | .30934 | 3.2327 | .95534 | **.80** | .71736 | 1.0296 | .97121 | .69671 |
| .31 | .30506 | .32033 | 3.1218 | .95233 | .81 | .72429 | 1.0505 | .95197 | .68950 |
| .32 | .31457 | .33139 | 3.0176 | .94924 | .82 | .73115 | 1.0717 | .93309 | .68222 |
| .33 | .32404 | .34252 | 2.9195 | .94604 | .83 | .73793 | 1.0934 | .91455 | .67488 |
| .34 | .33349 | .35374 | 2.8270 | .94275 | .84 | .74464 | 1.1156 | .89635 | .66746 |
| .35 | .34290 | .36503 | 2.7395 | .93937 | .85 | .75128 | 1.1383 | .87848 | .65998 |
| .36 | .35227 | .37640 | 2.6567 | .93590 | .86 | .75784 | 1.1616 | .86091 | .65244 |
| .37 | .36162 | .38786 | 2.5782 | .93233 | .87 | .76433 | 1.1853 | .84365 | .64483 |
| .38 | .37092 | .39941 | 2.5037 | .92866 | .88 | .77074 | 1.2097 | .82668 | .63715 |
| .39 | .38019 | .41105 | 2.4328 | .92491 | .89 | .77707 | 1.2346 | .80998 | .62941 |
| **.40** | .38942 | .42279 | 2.3652 | .92106 | **.90** | .78333 | 1.2602 | .79355 | .62161 |
| .41 | .39861 | .43463 | 2.3008 | .91712 | .91 | .78950 | 1.2864 | .77738 | .61375 |
| .42 | .40776 | .44657 | 2.2393 | .91309 | .92 | .79560 | 1.3133 | .76146 | .60582 |
| .43 | .41687 | .45862 | 2.1804 | .90897 | .93 | .80162 | 1.3409 | .74578 | .59783 |
| .44 | .42594 | .47078 | 2.1241 | .90475 | .94 | .80756 | 1.3692 | .73034 | .58979 |
| .45 | .43497 | .48306 | 2.0702 | .90045 | .95 | .81342 | 1.3984 | .71511 | .58168 |
| .46 | .44395 | .49545 | 2.1084 | .89605 | .96 | .81919 | 1.4284 | .70010 | .57352 |
| .47 | .45289 | .50797 | 1.9686 | .89157 | .97 | .82489 | 1.4592 | .68531 | .56530 |
| .48 | .46178 | .52061 | 1.9208 | .88699 | .98 | .83050 | 1.4910 | .67071 | .55702 |
| .49 | .47063 | .53339 | 1.8748 | .88233 | .99 | .83603 | 1.5237 | .65631 | .54869 |
| **.50** | .47943 | .54630 | 1.8305 | .87758 | **1.00** | .84147 | 1.5574 | .64209 | .54030 |
| $x$ | Sin $x$ | Tan $x$ | Cot $x$ | Cos $x$ | $x$ | Sin $x$ | Tan $x$ | Cot $x$ | Cos $x$ |

**Table V**
Trigonometric
Functions
of Real
Numbers
(Continued)

| x | Sin x | Tan x | Cot x | Cos x | x | Sin x | Tan x | Cot x | Cos x |
|---|---|---|---|---|---|---|---|---|---|
| **1.00** | .84147 | 1.5574 | .64209 | .54030 | **1.50** | .99749 | 14.101 | .07091 | .07074 |
| 1.01 | .84683 | 1.5922 | .62806 | .53186 | 1.51 | .99815 | 16.428 | .06087 | .06076 |
| 1.02 | .85211 | 1.6281 | .61420 | .52337 | 1.52 | .99871 | 19.670 | .05084 | .05077 |
| 1.03 | .85730 | 1.6652 | .60051 | .51482 | 1.53 | .99917 | 24.498 | .04082 | .04079 |
| 1.04 | .86240 | 1.7036 | .58699 | .50622 | 1.54 | .99953 | 32.461 | .03081 | .03079 |
| 1.05 | .86742 | 1.7433 | .57362 | .49757 | 1.55 | .99978 | 48.078 | .02080 | .02079 |
| 1.06 | .87236 | 1.7844 | .56040 | .48887 | 1.56 | .99994 | 92.621 | .01080 | .01080 |
| 1.07 | .87720 | 1.8270 | .54734 | .48012 | 1.57 | 1.00000 | 1255.8 | .00080 | .00080 |
| 1.08 | .88196 | 1.8712 | .53441 | .47133 | 1.58 | .99996 | −108.65 | −.00920 | −.00920 |
| 1.09 | .88663 | 1.9171 | .52162 | .46249 | 1.59 | .99982 | −52.067 | −.01921 | −.01920 |
| **1.10** | .89121 | 1.9648 | .50897 | .45360 | **1.60** | .99957 | −34.233 | −.02921 | −.02920 |
| 1.11 | .89570 | 2.0143 | .49644 | .44466 | 1.61 | .99923 | −25.495 | −.03922 | −.03919 |
| 1.12 | .90010 | 2.0660 | .48404 | .43568 | 1.62 | .99879 | −20.307 | −.04924 | −.04918 |
| 1.13 | .90441 | 2.1198 | .47175 | .42666 | 1.63 | .99825 | −16.871 | −.05927 | −.05917 |
| 1.14 | .90863 | 2.1759 | .45959 | .41759 | 1.64 | .99761 | −14.427 | −.06931 | −.06915 |
| 1.15 | .91276 | 2.2345 | .44753 | .40849 | 1.65 | .99687 | −12.599 | −.07937 | −.07912 |
| 1.16 | .91680 | 2.2958 | .43558 | .39934 | 1.66 | .99602 | −11.181 | −.08944 | −.08909 |
| 1.17 | .92075 | 2.3600 | .42373 | .39015 | 1.67 | .99508 | −10.047 | −.09953 | −.09904 |
| 1.18 | .92461 | 2.4273 | .41199 | .38092 | 1.68 | .99404 | − 9.1208 | −.10964 | −.10899 |
| 1.19 | .92837 | 2.4979 | .40034 | .37166 | 1.69 | .99290 | − 8.3492 | −.11977 | −.11892 |
| **1.20** | .93204 | 2.5722 | .38878 | .36236 | **1.70** | .99166 | − 7.6966 | −.12993 | −.12884 |
| 1.21 | .93562 | 2.6503 | .37731 | .35302 | 1.71 | .99033 | − 7.1373 | −.14011 | −.13875 |
| 1.22 | .93910 | 2.7328 | .36593 | .34365 | 1.72 | .98889 | − 6.6524 | −.15032 | −.14865 |
| 1.23 | .94249 | 2.8198 | .35463 | .33424 | 1.73 | .98735 | − 6.2281 | −.16056 | −.15853 |
| 1.24 | .94578 | 2.9119 | .34341 | .32480 | 1.74 | .98572 | − 5.8535 | −.17084 | −.16840 |
| 1.25 | .94898 | 3.0096 | .33227 | .31532 | 1.75 | .98399 | − 5.5204 | −.18115 | −.17825 |
| 1.26 | .95209 | 3.1133 | .32121 | .30582 | 1.76 | .98215 | − 5.2221 | −.19149 | −.18808 |
| 1.27 | .95510 | 3.2236 | .31021 | .29628 | 1.77 | .98022 | − 4.9534 | −.20188 | −.19789 |
| 1.28 | .95802 | 3.3413 | .29928 | .28672 | 1.78 | .97820 | − 4.7101 | −.21231 | −.20768 |
| 1.29 | .96084 | 3.4672 | .28842 | .27712 | 1.79 | .97607 | − 4.4887 | −.22278 | −.21745 |
| **1.30** | .96356 | 3.6021 | .27762 | .26750 | **1.80** | .97385 | − 4.2863 | −.23330 | −.22720 |
| 1.31 | .96618 | 3.7471 | .26687 | .25785 | 1.81 | .97153 | − 4.1005 | −.24387 | −.23693 |
| 1.32 | .96872 | 3.9033 | .25619 | .24818 | 1.82 | .96911 | − 3.9294 | −.25449 | −.24663 |
| 1.33 | .97115 | 4.0723 | .24556 | .23848 | 1.83 | .96659 | − 3.7712 | −.26517 | −.25631 |
| 1.34 | .97348 | 4.2556 | .23498 | .22875 | 1.84 | .96398 | − 3.6245 | −.27590 | −.26596 |
| 1.35 | .97572 | 4.4552 | .22446 | .21901 | 1.85 | .96128 | − 3.4881 | −.28669 | −.27559 |
| 1.36 | .97786 | 4.6734 | .21398 | .20924 | 1.86 | .95847 | − 3.3608 | −.29755 | −.28519 |
| 1.37 | .97991 | 4.9131 | .20354 | .19945 | 1.87 | .95557 | − 3.2419 | −.30846 | −.29476 |
| 1.38 | .98185 | 5.1774 | .19315 | .18964 | 1.88 | .95258 | − 3.1304 | −.31945 | −.30430 |
| 1.39 | .98370 | 5.4707 | .18279 | .17981 | 1.89 | .94949 | − 3.0257 | −.33051 | −.31381 |
| **1.40** | .98545 | 5.7979 | .17248 | .16997 | **1.90** | .94630 | − 2.9271 | −.34164 | −.32329 |
| 1.41 | .98710 | 6.1654 | .16220 | .16010 | 1.91 | .94302 | − 2.8341 | −.35284 | −.33274 |
| 1.42 | .98865 | 6.5811 | .15195 | .15023 | 1.92 | .93965 | − 2.7463 | −.36413 | −.34215 |
| 1.43 | .99010 | 7.0555 | .14173 | .14033 | 1.93 | .93618 | − 2.6632 | −.37549 | −.35153 |
| 1.44 | .99146 | 7.6018 | .13155 | .13042 | 1.94 | .93262 | − 2.5843 | −.38695 | −.36087 |
| 1.45 | .99271 | 8.2381 | .12139 | .12050 | 1.95 | .92896 | − 2.5095 | −.39849 | −.37018 |
| 1.46 | .99387 | 8.9886 | .11125 | .11057 | 1.96 | .92521 | − 2.4383 | −.41012 | −.37945 |
| 1.47 | .99492 | 9.8874 | .10114 | .10063 | 1.97 | .92137 | − 2.3705 | −.42185 | −.38868 |
| 1.48 | .99588 | 10.983 | .09105 | .09067 | 1.98 | .91744 | − 2.3058 | −.43368 | −.39788 |
| 1.49 | .99674 | 12.350 | .08097 | .08071 | 1.99 | .91341 | − 2.2441 | −.44562 | −.40703 |
| **1.50** | .99749 | 14.101 | .07081 | .07074 | **2.00** | .90930 | − 2.1850 | −.45766 | −.41615 |
| x | Sin x | Tan x | Cot x | Cos x | x | Sin x | Tan x | Cot x | Cos x |

**Table VI**
Trigonometric
Functions
of Angles

| Degrees | Radians | Sin | Cos | Tan | Cot | Sec | Csc | | |
|---|---|---|---|---|---|---|---|---|---|
| 0 | .0000 | .0000 | 1.0000 | .0000 | ∞ | 1.0000 | ∞ | 1.5708 | 90 |
| 1 | .0175 | .0175 | .9998 | .0175 | 57.2900 | 1.0002 | 57.299 | 1.5533 | 89 |
| 2 | .0349 | .0349 | .9994 | .0349 | 28.6363 | 1.0006 | 28.654 | 1.5359 | 88 |
| 3 | .0524 | .0523 | .9986 | .0524 | 19.0811 | 1.0014 | 19.107 | 1.5184 | 87 |
| 4 | .0698 | .0698 | .9976 | .0699 | 14.3007 | 1.0024 | 14.336 | 1.5010 | 86 |
| 5 | .0873 | .0872 | .9962 | .0875 | 11.4301 | 1.0038 | 11.474 | 1.4835 | 85 |
| 6 | .1047 | .1045 | .9945 | .1051 | 9.5144 | 1.0055 | 9.5668 | 1.4661 | 84 |
| 7 | .1222 | .1219 | .9925 | .1228 | 8.1443 | 1.0075 | 8.2055 | 1.4486 | 83 |
| 8 | .1396 | .1392 | .9903 | .1405 | 7.1154 | 1.0098 | 7.1853 | 1.4312 | 82 |
| 9 | .1571 | .1564 | .9877 | .1584 | 6.3138 | 1.0125 | 6.3925 | 1.4137 | 81 |
| 10 | .1745 | .1736 | .9848 | .1763 | 5.6713 | 1.0154 | 5.7588 | 1.3963 | 80 |
| 11 | .1920 | .1908 | .9816 | .1944 | 5.1446 | 1.0187 | 5.2408 | 1.3788 | 79 |
| 12 | .2094 | .2079 | .9781 | .2126 | 4.7046 | 1.0223 | 4.8097 | 1.3614 | 78 |
| 13 | .2269 | .2250 | .9744 | .2309 | 4.3315 | 1.0263 | 4.4454 | 1.3439 | 77 |
| 14 | .2443 | .2419 | .9703 | .2493 | 4.0108 | 1.0306 | 4.1336 | 1.3265 | 76 |
| 15 | .2618 | .2588 | .9659 | .2679 | 3.7321 | 1.0353 | 3.8637 | 1.3090 | 75 |
| 16 | .2793 | .2756 | .9613 | .2867 | 3.4874 | 1.0403 | 3.6280 | 1.2915 | 74 |
| 17 | .2967 | .2924 | .9563 | .3057 | 3.2709 | 1.0457 | 3.4203 | 1.2741 | 73 |
| 18 | .3142 | .3090 | .9511 | .3249 | 3.0777 | 1.0515 | 3.2361 | 1.2566 | 72 |
| 19 | .3316 | .3256 | .9455 | .3443 | 2.9042 | 1.0576 | 3.0716 | 1.2392 | 71 |
| 20 | .3491 | .3420 | .9397 | .3640 | 2.7475 | 1.0642 | 2.9238 | 1.2217 | 70 |
| 21 | .3665 | .3584 | .9336 | .3839 | 2.6051 | 1.0711 | 2.7904 | 1.2043 | 69 |
| 22 | .3840 | .3746 | .9272 | .4040 | 2.4751 | 1.0785 | 2.6695 | 1.1868 | 68 |
| 23 | .4014 | .3907 | .9205 | .4245 | 2.3559 | 1.0864 | 2.5593 | 1.1694 | 67 |
| 24 | .4189 | .4067 | .9135 | .4452 | 2.2460 | 1.0946 | 2.4586 | 1.1519 | 66 |
| 25 | .4363 | .4226 | .9063 | .4663 | 2.1445 | 1.1034 | 2.3662 | 1.1345 | 65 |
| 26 | .4538 | .4384 | .8988 | .4877 | 2.0503 | 1.1126 | 2.2812 | 1.1170 | 64 |
| 27 | .4712 | .4540 | .8910 | .5095 | 1.9626 | 1.1223 | 2.2027 | 1.0996 | 63 |
| 28 | .4887 | .4695 | .8829 | .5317 | 1.8807 | 1.1326 | 2.1301 | 1.0821 | 62 |
| 29 | .5061 | .4848 | .8746 | .5543 | 1.8040 | 1.1434 | 2.0627 | 1.0647 | 61 |
| 30 | .5236 | .5000 | .8660 | .5774 | 1.7321 | 1.1547 | 2.0000 | 1.0472 | 60 |
| 31 | .5411 | .5150 | .8572 | .6009 | 1.6643 | 1.1666 | 1.9416 | 1.0297 | 59 |
| 32 | .5585 | .5299 | .8480 | .6249 | 1.6003 | 1.1792 | 1.8871 | 1.0123 | 58 |
| 33 | .5760 | .5446 | .8387 | .6494 | 1.5399 | 1.1924 | 1.8361 | .9948 | 57 |
| 34 | .5934 | .5592 | .8290 | .6745 | 1.4826 | 1.2062 | 1.7883 | .9774 | 56 |
| 35 | .6109 | .5736 | .8192 | .7002 | 1.4281 | 1.2208 | 1.7434 | .9599 | 55 |
| 36 | .6283 | .5878 | .8090 | .7265 | 1.3764 | 1.2361 | 1.7013 | .9425 | 54 |
| 37 | .6458 | .6018 | .7986 | .7536 | 1.3270 | 1.2521 | 1.6616 | .9250 | 53 |
| 38 | .6632 | .6157 | .7880 | .7813 | 1.2799 | 1.2690 | 1.6243 | .9076 | 52 |
| 39 | .6807 | .6293 | .7771 | .8098 | 1.2349 | 1.2868 | 1.5890 | .8901 | 51 |
| 40 | .6981 | .6428 | .7660 | .8391 | 1.1918 | 1.3054 | 1.5557 | .8727 | 50 |
| 41 | .7156 | .6561 | .7547 | .8693 | 1.1504 | 1.3250 | 1.5243 | .8552 | 49 |
| 42 | .7330 | .6691 | .7431 | .9004 | 1.1106 | 1.3456 | 1.4945 | .8378 | 48 |
| 43 | .7505 | .6820 | .7314 | .9325 | 1.0724 | 1.3673 | 1.4663 | .8203 | 47 |
| 44 | .7679 | .6947 | .7193 | .9657 | 1.0355 | 1.3902 | 1.4396 | .8029 | 46 |
| 45 | .7854 | .7071 | .7071 | 1.0000 | 1.0000 | 1.4142 | 1.4142 | .7854 | 45 |
| | | Cos | Sin | Cot | Tan | Csc | Sec | Radians | Degrees |

**Table VII**
Squares, Cubes, Roots

| $n$ | $n^2$ | $\sqrt{n}$ | $n^3$ | $\sqrt[3]{n}$ | $n$ | $n^2$ | $\sqrt{n}$ | $n^3$ | $\sqrt[3]{n}$ |
|---|---|---|---|---|---|---|---|---|---|
| 1 | 1 | 1.000 | 1 | 1.000 | 51 | 2,601 | 7.141 | 132,651 | 3.708 |
| 2 | 4 | 1.414 | 8 | 1.260 | 52 | 2,704 | 7.211 | 140,608 | 3.733 |
| 3 | 9 | 1.732 | 27 | 1.442 | 53 | 2,809 | 7.280 | 148,877 | 3.756 |
| 4 | 16 | 2.000 | 64 | 1.587 | 54 | 2,916 | 7.348 | 157,464 | 3.780 |
| 5 | 25 | 2.236 | 125 | 1.710 | 55 | 3,025 | 7.416 | 166,375 | 3.803 |
| 6 | 36 | 2.449 | 216 | 1.817 | 56 | 3,136 | 7.483 | 175,616 | 3.826 |
| 7 | 49 | 2.646 | 343 | 1.913 | 57 | 3,249 | 7.550 | 185,193 | 3.849 |
| 8 | 64 | 2.828 | 512 | 2.000 | 58 | 3,364 | 7.616 | 195,112 | 3.871 |
| 9 | 81 | 3.000 | 729 | 2.080 | 59 | 3,481 | 7.681 | 205,379 | 3.893 |
| 10 | 100 | 3.162 | 1,000 | 2.154 | 60 | 3,600 | 7.746 | 216,000 | 3.915 |
| 11 | 121 | 3.317 | 1,331 | 2.224 | 61 | 3,721 | 7.810 | 226,981 | 3.936 |
| 12 | 144 | 3.464 | 1,728 | 2.289 | 62 | 3,844 | 7.874 | 238,328 | 3.958 |
| 13 | 169 | 3.606 | 2,197 | 2.351 | 63 | 3,969 | 7.937 | 250,047 | 3.979 |
| 14 | 196 | 3 742 | 2,744 | 2.410 | 64 | 4,096 | 8.000 | 262,144 | 4.000 |
| 15 | 225 | 3.873 | 3,375 | 2.466 | 65 | 4,225 | 8.062 | 274,625 | 4.021 |
| 16 | 256 | 4.000 | 4,096 | 2.520 | 66 | 4,356 | 8.124 | 287,496 | 4.041 |
| 17 | 289 | 4.123 | 4,913 | 2.571 | 67 | 4,489 | 8.185 | 300,763 | 4.062 |
| 18 | 324 | 4.243 | 5,832 | 2.621 | 68 | 4,624 | 8.246 | 314,432 | 4.082 |
| 19 | 361 | 4.359 | 6,859 | 2.668 | 69 | 4,761 | 8.307 | 328,509 | 4.102 |
| 20 | 400 | 4.472 | 8,000 | 2.714 | 70 | 4,900 | 8.367 | 343,000 | 4.121 |
| 21 | 441 | 4.583 | 9,261 | 2.759 | 71 | 5,041 | 8.426 | 357,911 | 4.141 |
| 22 | 484 | 4.690 | 10,648 | 2.802 | 72 | 5,184 | 8.485 | 373,248 | 4.160 |
| 23 | 529 | 4.796 | 12,167 | 2.844 | 73 | 5,329 | 8.544 | 389,017 | 4.179 |
| 24 | 576 | 4.899 | 13,824 | 2.884 | 74 | 5,476 | 8.602 | 405,224 | 4.198 |
| 25 | 625 | 5.000 | 15,625 | 2.924 | 75 | 5,625 | 8.660 | 421,875 | 4.217 |
| 26 | 676 | 5.099 | 17,576 | 2.962 | 76 | 5,776 | 8.718 | 438,976 | 4.236 |
| 27 | 729 | 5.196 | 19,683 | 3.000 | 77 | 5,929 | 8.775 | 456,533 | 4.254 |
| 28 | 784 | 5.292 | 21,952 | 3.037 | 78 | 6,084 | 8.832 | 474,552 | 4.273 |
| 29 | 841 | 5.385 | 24,389 | 3.072 | 79 | 6,241 | 8.888 | 493,039 | 4.291 |
| 30 | 900 | 5.477 | 27,000 | 3.107 | 80 | 6,400 | 8.944 | 512,000 | 4.309 |
| 31 | 961 | 5.568 | 29,791 | 3.141 | 81 | 6,561 | 9.000 | 531,441 | 4.327 |
| 32 | 1,024 | 5.657 | 32,768 | 3.175 | 82 | 6,724 | 9.055 | 551,368 | 4.344 |
| 33 | 1,089 | 5.745 | 35,937 | 3.208 | 83 | 6,889 | 9.110 | 571,787 | 4.362 |
| 34 | 1,156 | 5.831 | 39,304 | 3.240 | 84 | 7,056 | 9.165 | 592,704 | 4.380 |
| 35 | 1,225 | 5.916 | 42,875 | 3.271 | 85 | 7,225 | 9.220 | 614,125 | 4.397 |
| 36 | 1,296 | 6.000 | 46,656 | 3.302 | 86 | 7,396 | 9.274 | 636,056 | 4.414 |
| 37 | 1,369 | 6.083 | 50,653 | 3.332 | 87 | 7,569 | 9.327 | 658,503 | 4.431 |
| 38 | 1,444 | 6.164 | 54,872 | 3.362 | 88 | 7,744 | 9.381 | 681,472 | 4.448 |
| 39 | 1,521 | 6.245 | 59,319 | 3.391 | 89 | 7,921 | 9.434 | 704,969 | 4.465 |
| 40 | 1,600 | 6.325 | 64,000 | 3.420 | 90 | 8,100 | 9.487 | 729,000 | 4.481 |
| 41 | 1,681 | 6.403 | 68,921 | 3.448 | 91 | 8,281 | 9.539 | 753,571 | 4.498 |
| 42 | 1,764 | 6.481 | 74,088 | 3.476 | 92 | 8,464 | 9.592 | 778,688 | 4.514 |
| 43 | 1,849 | 6.557 | 79,507 | 3.503 | 93 | 8,649 | 9.644 | 804,357 | 4.531 |
| 44 | 1,936 | 6.633 | 85,184 | 3.530 | 94 | 8,836 | 9.695 | 830,584 | 4.547 |
| 45 | 2,025 | 6.708 | 91,125 | 3.557 | 95 | 9,025 | 9.747 | 857,375 | 4.563 |
| 46 | 2,116 | 6.782 | 97,336 | 3.583 | 96 | 9,216 | 9.798 | 884,736 | 4.579 |
| 47 | 2,209 | 6.856 | 103,823 | 3.609 | 97 | 9,409 | 9.849 | 912,673 | 4.595 |
| 48 | 2,304 | 6.928 | 110,592 | 3.634 | 98 | 9,604 | 9.899 | 941,192 | 4.610 |
| 49 | 2,401 | 7.000 | 117,649 | 3.659 | 99 | 9,801 | 9.950 | 970,299 | 4.626 |
| 50 | 2,500 | 7.071 | 125,000 | 3.684 | 100 | 10,000 | 10.000 | 1,000,000 | 4.642 |
| $n$ | $n^2$ | $\sqrt{n}$ | $n^3$ | $\sqrt[3]{n}$ | $n$ | $n^2$ | $\sqrt{n}$ | $n^3$ | $\sqrt[3]{n}$ |

# ANSWERS TO ODD-NUMBERED PROBLEMS

**PROBLEMS 1.6**

**1** See Fig. A.1.

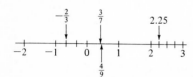

**Figure A.1**

**3** **a** 4 **b** $\frac{11}{16}$ **c** 0 **d** no number

**5** **a** $\frac{2}{13} = 0.\overline{153846}$ **b** $\frac{5}{13} = 0.\overline{384615}$ **c** $\frac{2}{39} = 0.\overline{051282}$
**d** $\frac{7}{39} = 0.\overline{179487}$

**7** **a** $0.\overline{549} = \frac{549}{999}$ **b** $0.0\overline{15} = \frac{1}{66}$ **c** $0.0\overline{27} = \frac{1}{37}$ **d** $1.\overline{81} = \frac{20}{11}$

**9** Let $a/b$ be smaller than $c/d$. We must prove that the length of the segment from $a/b$ to the mean is the same as the length of the segment from the mean to $c/d$, or

$$\frac{\frac{a}{b} + \frac{c}{d}}{2} - \frac{a}{b} = \frac{c}{d} - \frac{\frac{a}{b} + \frac{c}{d}}{2}$$

Add $a/b$ to both sides. Then

$$\frac{\frac{a}{b}+\frac{c}{d}}{2}=\left(\frac{a}{b}+\frac{c}{d}\right)-\left(\frac{\frac{a}{b}+\frac{c}{d}}{2}\right)$$

which is obviously true.

## PROBLEMS 1.11

**1 a** $\frac{\frac{2}{3}}{\frac{8}{7}}=(\frac{2}{3})(\frac{7}{8})=\frac{7}{12}$  **b** $\frac{\frac{-9}{4}}{\frac{3}{2}}=(\frac{-9}{4})(\frac{2}{3})=\frac{-3}{2}$  **c** $\frac{\frac{5}{6}}{\frac{-6}{5}}=(\frac{5}{6})(\frac{-5}{6})=\frac{-25}{36}$

   **d** $\frac{\frac{-1}{2}}{\frac{-10}{11}}=(\frac{1}{2})(\frac{11}{10})=\frac{11}{20}$

**3 a** No. $\sqrt{2}+(-\sqrt{2})=0$ (rational)  **b** No. $\sqrt{2}\times\sqrt{2}=2$ (rational)

**5 a** 1.74  **b** 1.7574  **c** 1.757574  **d** $1.\overline{75}$. Proof: $0.\overline{87}=\frac{87}{99}$

   $2(\frac{87}{99})=\frac{174}{99}=1.\overline{75}$

**7** 37,388/99,900 (can be reduced but no need to)

**9 a** $-4,4$  **b** no number  **c** $-1,7$  **d** $-7,1$  **e** $-\frac{1}{2}$

   **f** $\pm1,\pm\sqrt{3}$

**11 a** $3^2/2^6$  **b** $2^6\cdot3^4$  **c** $1/(2^{1/6}\cdot3^{10/3})$  **d** $2^{19/2}\cdot3^3$

**13 a** $\frac{1}{4}-\frac{1}{4}\sqrt{5}$  **b** $11-6\sqrt{3}$  **c** $-\frac{23}{31}-\frac{30}{31}\sqrt{2}$  **d** $2-\frac{1}{3}\sqrt{6}$

## PROBLEMS 1.12

**1** Cannot divide $x(x-1)=0$ by $x$ to yield $x-1=0$, because $x=0$ and we never divide by 0.

## PROBLEMS 2.3

**1 a** 2  **b** $-1$  **c** $\frac{19}{7}$  **d** $\frac{1}{3}$  **e** 5  **f** 26

   **g** $\frac{9}{2}$  **h** none

**3 a** $1,-5$  **b** $\frac{1}{2},-3$  **c** none  **d** $-\frac{1}{3},\frac{1}{2}$  **e** $1,-7$

   **f** $-\frac{5}{2},3$  **g** 5  **h** none

**5 a** $(x+7)(x+2)=0$  **b** $(x+2)(x-11)=0$  **c** $(x-\frac{2}{3})(x+6)=0$

   **d** $(x-\frac{1}{3})(x+\frac{5}{2})=0$

## PROBLEMS 2.4

**1 a** $x\leq4$  **b** $x\geq-6$  **c** $x>2$  **d** $x<-10$  **e** $x\geq-\frac{3}{5}$

   **f** $x\geq\frac{5}{3}$  See Fig. A.2.

**3 a** $1\leq x\leq2$  **b** $x\leq-1$ or $x\geq2$  **c** $x<-3$ or $x>\frac{1}{2}$

   **d** $-2<x<1$  See Fig. A.3.

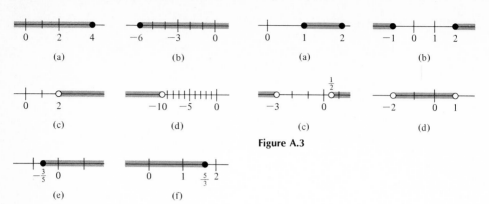

(a)

(b)

(c)

(d)

(e)

(f)

**Figure A.2**

**Figure A.3**

## PROBLEMS 2.5

**1** **a** $-5 \leq x \leq 5$ **b** $x \leq -2$ or $x \geq 2$ **c** $x \leq -4$ or $x \geq 4$
**d** $-1 \leq x \leq 1$ **e** $3 < x < 7$ **f** $-7 \leq x \leq -1$ **g** no number
**h** all numbers See Fig. A.4.

**3** **a** $-\frac{4}{3} \leq x \leq 3$ **b** $-3 \leq x \leq 3$ **c** all numbers
**d** $-\frac{1}{5} < x < 1$ **e** $-2 < x < 2$ **f** $x \leq -1$ or $x \geq \frac{1}{3}$
See Fig. A.5.

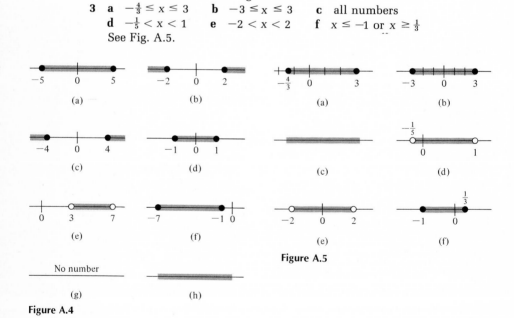

(a)

(b)

(c)

(d)

(e)

(f)

No number

(g)

(h)

**Figure A.4**

**Figure A.5**

## PROBLEMS 3.1

**1** $\frac{4}{3}$, 1, 0, $f(4)$ does not exist **3** $\frac{2}{3}$, $\frac{3}{7}$, $\frac{4}{13}$, $\frac{5}{21}$ **5** 68, 4, 148

**7** $\sqrt{2 + \sqrt{2}}$, 2 **9** Yes; $\theta$

## PROBLEMS 3.2

**1** Domain, reals. Range, reals
**3** Domain, $\{x \mid x \geq -\frac{3}{2}\}$. Range, $\{y \mid y \geq 0\}$

## PROBLEMS 3.3

**1** The table. Either. $\{1, 2, 3\}$, $\{1, 2, 3\}$
**3** See Fig. A.6. Domain, reals. Range, $\{-1, 0, 1\}$
**5** The graph. Domain, $\{x \mid -2 \leq x \leq 1\}$. Range, $\{y \mid 0 \leq y \leq 1\}$
**7** $y = x^2 - 2x^3$. Domain, reals. Range, reals

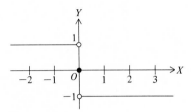

**Figure A.6**

## PROBLEMS 3.4

**1** **a** $d_f =$ reals; $d_g =$ reals except 0   **b** $f(x) \pm g(x) = 2x - 2 \pm 3/x$;
$d_{f \pm g} =$ reals except 0   **c** $f(x) \cdot g(x) = (6x - 6)/x$; $d_{fg} =$ reals except 0

**d** $f(x)/g(x) = \dfrac{2x - 2}{(3/x)}$; $d_{f/g} =$ reals except 0

$= \dfrac{2x^2 - 2x}{3}$ for all $x$ except 0

**3** **a** $d_f = \{x \mid x \leq 1\}$; $d_g =$ reals except 4   **b** $f(x) \pm g(x) =$
$\sqrt{1-x} \pm 3/(x-4)$; $d_{f \pm g} = d_f \cap d_g = \{x \mid x \leq 1\}$   **c** $f(x) \cdot g(x) =$
$3\sqrt{1-x}/(x-4)$; $d_{fg} = d_f \cap d_g = \{x \mid x \leq 1\}$   **d** $f(x)/g(x) =$
$\dfrac{\sqrt{1-x}}{3/(x-4)}$; $d_{f/g} = d_f \cap d_g = \{x \mid x \leq 1\}$

**5** **a** $d_f =$ reals; $d_g =$ reals   **b** $f(x) \pm g(x) = x + 5 \pm x(x + 5)$;
$d_{f+g} =$ reals   **c** $f(x) \cdot g(x) = x(x + 5)^2$; $d_{fg} =$ reals   **d** $f(x) \cdot g(x) =$
$(x + 5)/x(x + 5)$; $d_{f/g} =$ reals except $-5$, $0 = 1/x$ except for $x = -5$, 0

**7** **a** $d_f = \{x \mid x \leq 2\}$; $d_g = \{x \mid x \geq 4\}$   **b–d** Since $d_f \cap d_g$ is null, the operations do not apply.

## PROBLEMS 3.5

**1** $f(g(x)) = |x^2 - 3x + 1|$; $d_{f \cdot g} =$ reals
**3** $f(g(x)) = (x^2 + x^{-2})^3 - (x^2 + x^{-2}) + 6$; $d_{f \cdot g} =$ reals except 0

**5** $f(g(x)) = \dfrac{1}{1 + \dfrac{1}{1 + x}}$; $d_{f \cdot g}$ = reals except $-1, -2$

$\qquad = (1 + x)/(2 + x)$     except $-1, -2$

**7** $f(g(x)) = \sqrt{x^2} = |x|$; $d_{f \cdot g}$ = reals

**9** $f(f(x)) = 2(2x - x^2) - (2x - x^2)^2 = -x^4 + 4x^3 - 6x^2 + 4x$

**11** $C(s) = 6s$, $s = 12m$, $C(m) = 72m$

## PROBLEMS 4.1

**1**   Function; either $t$ or $v$     **3**   See Fig. A.7; $AC = 50$

**5**   $D(-1, 0)$     **7**   $D(3, 0)$

**9**   If $(b, 0)$ is to the right of $(a, 0)$ on the number line, then the length is $b - a = |b - a|$. If $(b, 0)$ is to the left of $(a, 0)$, then the length is $a - b = |a - b| = |b - a|$.

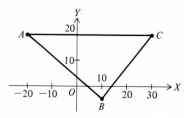

**Figure A.7**

## PROBLEMS 4.2

**1**   See Fig. A.8.     **3**   See Fig. A.9.     **5**   See Fig. A.10.

**7**   See Fig. A.11.     **9**   See Fig. A.12.     **11**   See Fig. A.13.

## PROBLEMS 4.3

**1**   $8x + 3y - 14 = 0$     **3**   $y = 6$

**5**   $10x - y - 27 = 0$     **7**   $x = 6$

**9**   $y = -3$     **11**   $y = 2x - 13$

**13**   $5x - 3y + 4 = 0$

## PROBLEMS 4.4

**1**   See Fig. A.14.     **3**   See Fig. A.15.     **5**   See Fig. A.16.

**7**   See Fig. A.17.     **9**   See Fig. A.18.     **11**   See Fig. A.19.

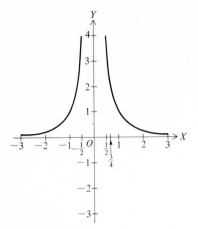

**Figure A.8**

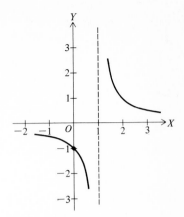

**Figure A.9**

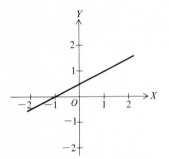

**Figure A.10**

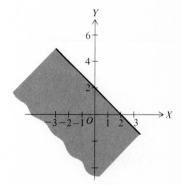

**Figure A.11**

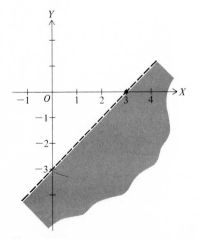

**Figure A.12**

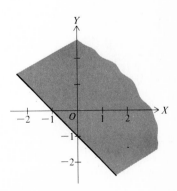

**Figure A.13**

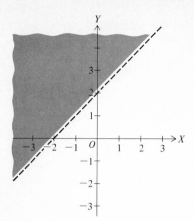

Figure A.14

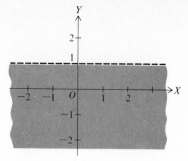

Figure A.15

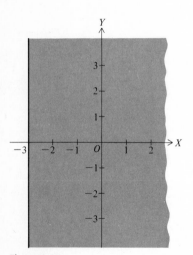

Figure A.16

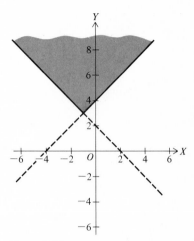

Figure A.17

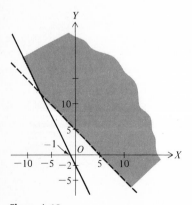

Figure A.18

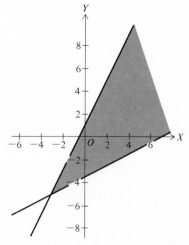

Figure A.19

13  See Fig. A.20.    15  See Fig. A.21.    17  See Fig. A.22.

19  See Fig. A.23.

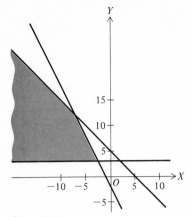

Figure A.20

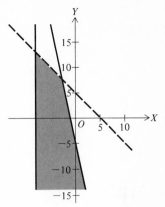

Figure A.21

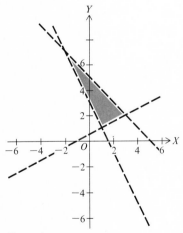

Figure A.22

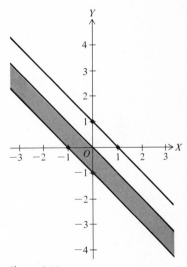

Figure A.23

## PROBLEMS 4.5

1  $\sqrt{89}$    3  $\sqrt{98}$

5  $\sqrt{18}$    7  18

9  $\overline{AB}^2 = 5$, $\overline{BC}^2 = 5$, $\overline{AC}^2 = 10$.   Angle $B$ is $90°$.

11  $\overline{AC}^2 = a^2 + b^2$.   $\overline{BD}^2 = a^2 + b^2$

## PROBLEMS 4.6

**1** $(x - 2)^2 + (y - 5)^2 = 36$    **3** $x^2 + (y + 8)^2 = 1$

**5** $(x - 2)^2 + (y - 2)^2 = 4$    **7** $(x - 1)^2 + (y - 1)^2 = 2$

**9** See Fig. A.24.    **11** See Fig. A.25.

**13** See Fig. A.26.    **15** See Fig. A.27.

**17** See Fig. A.28.    **19** $(-4, -6), (4, 2)$

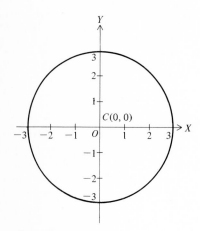

**Figure A.24**

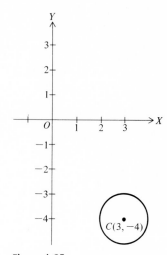

**Figure A.25**

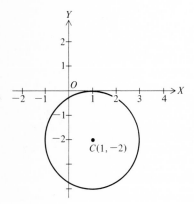

**Figure A.26**

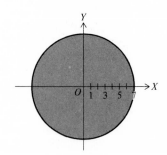

**Figure A.27**

## PROBLEMS 4.7

**1** Ellipse, Fig. A.29.

**3** Ellipse, Fig. A.30.

**5** Hyperbola; asymptotes: $3x \pm 5y = 0$; see Fig. A.31.

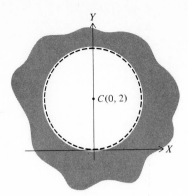

Figure A.28

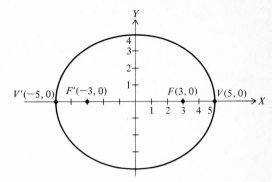

Figure A.29

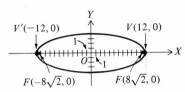

Figure A.30

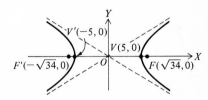

Figure A.31

**7** Hyperbola; asymptotes: $4x \pm 7y = 0$; see Fig. A.32.
**9** Parabola; directrix: $x = -3$; see Fig. A.33.
**11** Parabola; directrix: $x = 5$; see Fig. A.34.
**13** Parabola; see Fig. A.35.
**15** Parabola; see Fig. A.36.

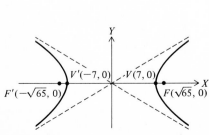

Figure A.32

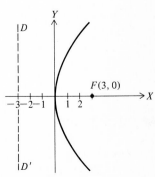

Figure A.33

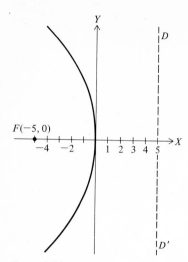

Figure A.34

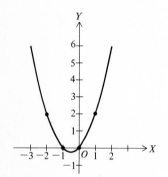

Figure A.35

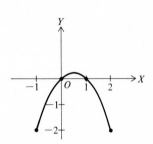

Figure A.36

## PROBLEMS 4.8

**1** $2x + y \leq 100$, $P = 10x + 10y$, $3x + 4y \leq 300$, $P(\max) = \$800$. See Fig. A.37.

**3** $0 \leq x \leq 55$, $x + 2y \leq 100$, $P = 2x + 3y$, $0 \leq y \leq 40$, $3x + y \leq 180$, $P(\max) = \$176$. See Fig. A.38.

**5** (N) $6x + 6y \geq 108$, $C = 5x + 10y$, (P) $2x + 9y \geq 85$, $C(\min) = \$150$, (L) $8x + 15y \geq 235$. See Fig. A.39.

## PROBLEMS 5.1

**1** **a** $2 - \sqrt{2}$ **b** $\sqrt{2}$ **c** $4\sqrt{2} - 4$ **d** $2 - 2(\sqrt[3]{2})^2 + 3(\sqrt[3]{2}) - \sqrt{2}$
**e** $-6 - \sqrt{2}$ **f** $6 - \sqrt{2}$

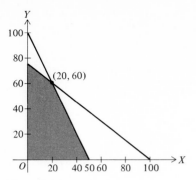

**Figure A.37**

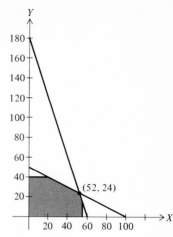

**Figure A.38**

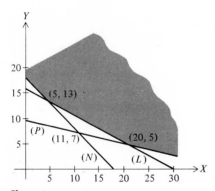

**Figure A.39**

**3 a** Polynomial, 10   **b** no   **c** no   **d** polynomial, 2   **e** no
**f** polynomial, 1000   **g** polynomial, 0
**5** 150,000, yes

## PROBLEMS 5.2

**1 a** $\frac{2}{9}$   **b** $\pi$   **c** 1, 2   **d** 1, 1   **e** $(-1 \pm \sqrt{5})/2$
**3 a** 1, −1   **b** 1, 2, 3, 4, 5   **c** $\pm 1, \pm \sqrt{2}$

## PROBLEMS 5.3

**1** $x^2 + 3x + 1$; $x^2 + x - 7$; $x^3 + 6x^2 + 5x - 12$; $x - 2 + 5/(x + 4)$
**3** $x^4 + 3x^2$; $x^4 - x^2 - 2$; $2x^6 + 3x^4 - x^2 - 1$; $\frac{1}{2}x^2 + \frac{1}{4} - \frac{5}{4}/(2x^2 + 1)$

5   $x^3 - 2x^2 + 8x + 2$; $x^3 + 2x^2 + 6x - 4$; $x^4 + x^3 + x^2 + 20x - 3$;
$x^2 - 5x + 22 - 67/(x + 3)$

7   $x^4 + 3x + 4$; $x^4 - x - 2$; $2x^5 + 3x^4 + 2x^2 + 5x + 3$;
$\frac{1}{2}x^3 - \frac{3}{4}x^2 + \frac{9}{8}x - \frac{19}{16} + \frac{73}{16}[1/(2x + 3)]$

## PROBLEMS 5.4

1  a  97     b  8445
3  a  $(-3 \pm \sqrt{5})/2$     b  $1, (-3 \pm \sqrt{5})/2$     c  $1, 2, -\frac{3}{2}$
   d  $1, 2, -\frac{3}{2}, -1$     e  $\pm\sqrt{2}, (-3 \pm \sqrt{5})/2$
5  $(x + 1)(x + 2)(x + 3)$

## PROBLEMS 5.5

1  a  $(x - 9)(x - 4)$     b  $(3x + 2)(x + 1)$     c  $\left(x + \dfrac{7 - \sqrt{5}}{2}\right)\left(x + \dfrac{7 + \sqrt{5}}{2}\right)$

   d  $(3x + 4)(3x - 4)$     e  $3\left(x - \dfrac{5 - \sqrt{13}}{6}\right)\left(x - \dfrac{5 + \sqrt{13}}{6}\right)$

3  $(x - 1)(x - 1)(x + 3)$

## PROBLEMS 5.6

1  See Fig. A.40.     3  See Fig. A.41.
5  See Fig. A.42.

## PROBLEMS 5.7

1  $-2.615$; error, $-0.04$     3  $2.239$; larger than 0

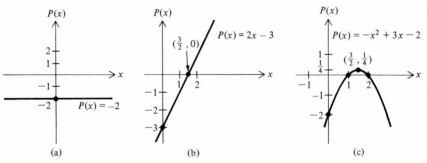

(a)                    (b)                    (c)

**Figure A.40**

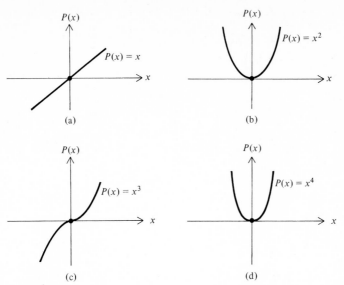

Figure A.41

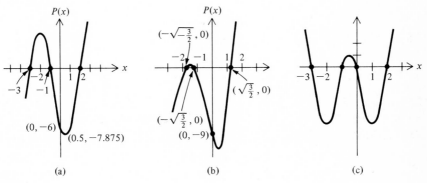

Figure A.42

## PROBLEMS 6.1

**1**

| x | $-2$ | $-1$ | $-\frac{1}{2}$ | $-\frac{1}{3}$ | 0 | $\frac{1}{3}$ | $\frac{1}{2}$ | 1 | 2 |
|---|---|---|---|---|---|---|---|---|---|
| $x^2$ | 4 | 1 | $\frac{1}{4}$ | $\frac{1}{9}$ | 0 | $\frac{1}{9}$ | $\frac{1}{4}$ | 1 | 4 |

**3**

| x | $-2$ | $-1$ | $-\frac{1}{2}$ | $-\frac{1}{3}$ | 0 | $\frac{1}{3}$ | $\frac{1}{2}$ | 1 | 2 |
|---|---|---|---|---|---|---|---|---|---|
| $2^{-x}$ | 4 | 2 | 1.414 | 1.260 | 1 | 0.794 | 0.707 | $\frac{1}{2}$ | $\frac{1}{4}$ |

**5** See Fig. A.43.    **7** See Fig. A.44.

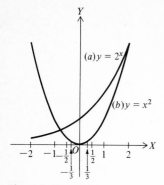

**Figure A.43**

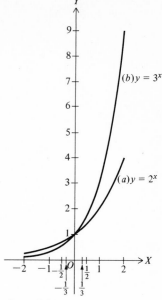

**Figure A.44**

## PROBLEMS 6.2

| | | | | | |
|---|---|---|---|---|---|
| **1** | $2^{1/2}$ | **3** | $8^{4.5}$ | **5** | $2^{x+1}$ |
| **7** | $3^{3x}$ | **9** | $2^{3x+3}$ | **11** | Cannot be simplified |
| **13** | Cannot be simplified | **15** | $a^{-x/2}$ | **17** | $b^{5x/6}$ |
| **19** | $c^{-3x/2}$ | | | | |

## PROBLEMS 6.3

| | | | |
|---|---|---|---|
| **1** | 2.7183 | **3** | 0.12246 |
| **5** | 2.0138 | **7** | 0.09072 |
| **9** | 7.2538 | **11** | Not an exponential function |
| **13** | An exponential function | **15** | Not an exponential function |
| **17** | Not an exponential function | **19** | An exponential function |
| **21** | See Fig. A.45. | **23** | See Fig. A.46. |
| **25** | See Fig. A.47. | **27** | See Fig. A.48. |
| **29** | See Fig. A.49. | **31** | See Fig. A.50. |
| **33** | See Fig. A.51. | | |

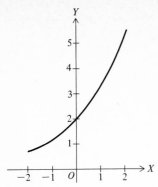

**Figure A.45**

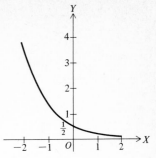

**Figure A.46**

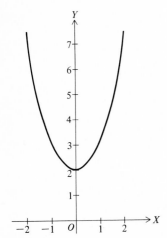

**Figure A.47**

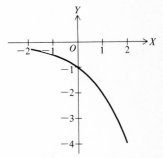

**Figure A.48**

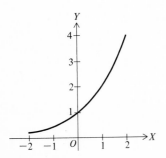

**Figure A.49**

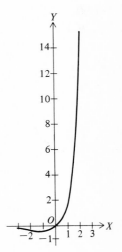

**Figure A.50**

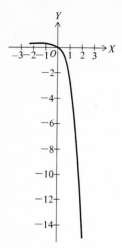

**Figure A.51**

## PROBLEMS 6.4

1  $A = (1 + \frac{0.04}{4})^2 = \$1.02$
3  $P = 100/(1 + \frac{0.04}{4}) = \$99.01$
5  $t = 11.7$ years
7  $A = (1 + \frac{0.05}{360})^{36,000} = \$148.36$ (by calculator)
9  $P = \$480.77$
11  30 percent
13  $\frac{1}{2} = e^{-0.038t}$; 18.42 centuries

## PROBLEMS 7.1

1  **a** One-to-one    **b** no, $f(-2) = f(2)$    **c** one-to-one
   **d** no, $f(\pi) = f(7)$
3  **a** One-to-one    **b** no, $f(-x) = f(x)$    **c** one-to-one
   **d** no, all are the same

## PROBLEMS 7.2

1  **a**

| x | 5 | 4 | 3 | 2 | 1 |
|---|---|---|---|---|---|
| $f^{-1}(x)$ | 1 | 2 | 3 | 4 | 5 |

   **c**

| x | −1 | 0 | 1 | 2 | 3 |
|---|----|---|---|---|---|
| $f^{-1}(x)$ | 1 | 7 | 8 | 16 | 100 |

**3  a** $f^{-1}(x) = -3x$    **c** $f^{-1}(x) = +\sqrt{x},\ x \geq 0$

## PROBLEMS 7.3

**1  a** $f^{-1}(x) = \sqrt[3]{\dfrac{x-4}{2}}$    **b** $f^{-1}(x) = \dfrac{4-x}{3}$    **c** $f^{-1}(x) = \dfrac{1}{x} - 3$

  **d**  not possible

**3**  See Fig. A.52.

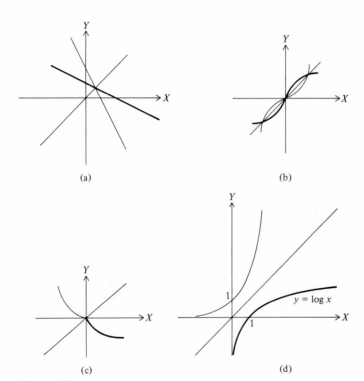

(a)

(b)

(c)

(d)

**Figure A.52**

$y = \log x$

## PROBLEMS 7.4

**1**  $h(x) = (26 - 2x)/9$

## PROBLEMS 8.1

**1  a**  $-2$    **b**  $0$    **c**  $\frac{1}{2}$    **d**  $\frac{1}{4}$

**3  a**  $-2$    **b**  $-1$    **c**  $2$    **d**  $-4$    **e**  $4$

**5 a** There is no x such that $1^x = 2$. **b** There is no x such that $10^x = 0$.

**7 a**

| x | 0.01 | 0.1 | 1 | $\sqrt{10}$ | 10 | $10^{3/2}$ | 100 |
|---|---|---|---|---|---|---|---|
| $\log_{10} x$ | $-2$ | $-1$ | 0 | $\frac{1}{2}$ | $\frac{1}{2}$ | $\frac{3}{2}$ | 2 |

**b**

| x | $\frac{1}{27}$ | $\frac{1}{9}$ | $\frac{1}{3}$ | 1 | $\sqrt{3}$ | 3 | 9 | 27 |
|---|---|---|---|---|---|---|---|---|
| $\log_{1/3} x$ | 3 | 2 | 1 | 0 | $-\frac{1}{2}$ | $-1$ | $-2$ | $-3$ |

## PROBLEMS 8.2

**1 a** $-5$ **b** $-2$ **c** 2
**5 a** $2.732 \times 10$ **b** $7.61 \times 10^{-3}$ **c** $4.9326511 \times 10^6$
  **d** $2.0 \times 10^{-1}$ **e** $4.713 \times 10^0$ **f** $1.7 \times 10$
**7 a** $-1$ **b** $2 + \log 8.936$ **c** $-3 + \log 3.72$
**9 a** 0.001 **b** 100 **c** 1
**11 a** $x = 2$ **b** $x = 4$, $x = -2$ is not in domain of $\log_2$ **c** $x = 1$

## PROBLEMS 8.3

**1 a** 0.5453 **b** $2 + 0.9786$ **c** $-3 + 0.2430$ **d** $1 + 0.6570$
**3 a** 4.04 **b** 8.512 **c** 805.2 **d** 0.00253

## PROBLEMS 8.4

**1 a** 0.5286 **b** 0.2493 **c** 0.685
**3 a** 7.944 **b** $-0.1801$
**5** About 11 years
**7** About 1.3 hours

## PROBLEMS 9.2

**1** See Fig. A.53. **3** See Fig. A.54. **5** See Fig. A.55.
**7** See Fig. A.56. **9** See Fig. A.57. **11 a** $\pi/2$

Figure A.53

Figure A.54

Figure A.55

Figure A.56

Figure A.57

**b** $\pi$   **c** $3\pi/2$   **d** $-\pi/2$   **e** $-\pi$   **f** $7\pi/2$
**g** $5\pi/2$   **h** $5\pi$   **i** $-9\pi/2$   **j** 0

## PROBLEMS 9.4

| arc S | 0 | $\pi/2$ | $\pi$ | $3\pi/2$ | $2\pi$ | $5\pi/2$ | $3\pi$ | $7\pi/2$ | $4\pi$ |
|---|---|---|---|---|---|---|---|---|---|
| **1** sin S | 0 | 1 | 0 | −1 | 0 | 1 | 0 | −1 | 0 |
| **3** tan S | 0 | — | 0 | — | 0 | — | 0 | — | 0 |

| arc S | 0 | $-\pi/2$ | $-\pi$ | $-3\pi/2$ | $-2\pi$ |
|---|---|---|---|---|---|
| **5** cos S | 1 | 0 | −1 | 0 | 1 |

## PROBLEMS 9.5

**1**

| arc S | $\pi/6$ | $5\pi/6$ | $7\pi/6$ | $11\pi/6$ |
|---|---|---|---|---|
| sin S | $\frac{1}{2}$ | $\frac{1}{2}$ | $-\frac{1}{2}$ | $-\frac{1}{2}$ |
| cos S | $\sqrt{3}/2$ | $-\sqrt{3}/2$ | $-\sqrt{3}/2$ | $\sqrt{3}/2$ |
| tan S | $1/\sqrt{3}$ | $-1/\sqrt{3}$ | $1/\sqrt{3}$ | $-1/\sqrt{3}$ |

**3**

| arc S | $\pi/4$ | $3\pi/4$ | $5\pi/4$ | $7\pi/4$ |
|---|---|---|---|---|
| sin S | $\sqrt{2}/2$ | $\sqrt{2}/2$ | $-\sqrt{2}/2$ | $-\sqrt{2}/2$ |
| cos S | $\sqrt{2}/2$ | $-\sqrt{2}/2$ | $-\sqrt{2}/2$ | $\sqrt{2}/2$ |
| tan S | 1 | −1 | 1 | −1 |

**5**

| arc S | 0 | $\pi/6$ | $\pi/4$ | $\pi/3$ | $\pi/2$ | $2\pi/3$ | $3\pi/4$ | $5\pi/6$ | $\pi$ |
|---|---|---|---|---|---|---|---|---|---|
| cos S | 1 | $\sqrt{3}/2$ | $\sqrt{2}/2$ | $\frac{1}{2}$ | 0 | $-\frac{1}{2}$ | $-\sqrt{2}/2$ | $-\sqrt{3}/2$ | −1 |

## PROBLEMS 9.6

**1** See Fig. A.58.   **3** See Fig. A.59.   **5** See Fig. A.60.
**7** See Fig. A.61.   **9** See Fig. A.62.   **11** See Fig. A.63.
**13** See Fig. A.64.   **15** See Fig. A.65.   **17** See Fig. A.66.

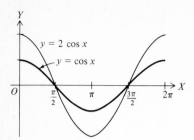

**Figure A.58**

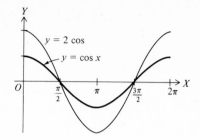

**Figure A.59**

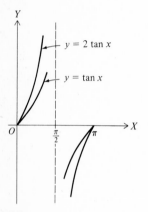

**Figure A.60**

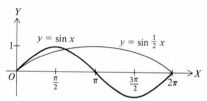

**Figure A.61**

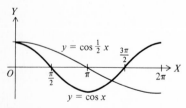

**Figure A.62**

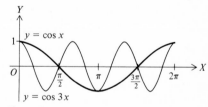

**Figure A.63**

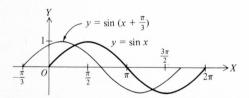

**Figure A.64**

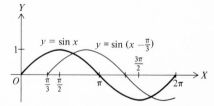

**Figure A.65**

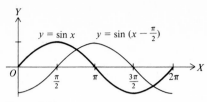

**Figure A.66**

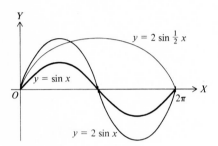

**Figure A.67**

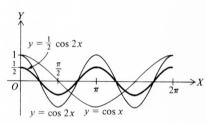

**Figure A.68**

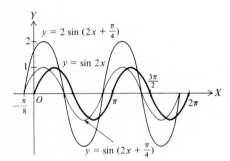

**Figure A.69**

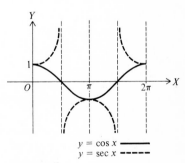

**Figure A.70**

**19**  See Fig. A.67.     **21**  See Fig. A.68.     **23**  See Fig. A.69.

## PROBLEMS 9.7

**1**  $\sin x = \sqrt{1 - \cos^2 x}$     $\tan x = \sqrt{1 - \cos^2 x}/\cos x$
$\csc x = 1/\sqrt{1 - \cos^2 x}$     $\sec x = 1/\cos x$     $\cot x = \cos x/\sqrt{1 - \cos^2 x}$
**3**  See Fig. A.70.

## PROBLEMS 9.8

| 9 | | sin | cos | tan |
|---|---|-----|-----|-----|
| | a | $\sqrt{\dfrac{1-\sqrt{2}/2}{2}}$ | $\sqrt{\dfrac{1+\sqrt{2}/2}{2}}$ | $\sqrt{\dfrac{2-\sqrt{2}}{2+\sqrt{2}}}$ |
| | | $=\frac{1}{2}\sqrt{2-\sqrt{2}}$ | $=\frac{1}{2}\sqrt{2+\sqrt{2}}$ | |
| | c | $\frac{1}{2}\sqrt{2+\sqrt{2}}$ | $-\frac{1}{2}\sqrt{2-\sqrt{2}}$ | $-\sqrt{\dfrac{2+\sqrt{2}}{2-\sqrt{2}}}$ |
| | e | $-\frac{1}{2}\sqrt{2-\sqrt{2}}$ | $-\frac{1}{2}\sqrt{2+\sqrt{2}}$ | $\sqrt{\dfrac{2-\sqrt{2}}{2+\sqrt{2}}}$ |
| | g | $-\frac{1}{2}\sqrt{2+\sqrt{2}}$ | $\frac{1}{2}\sqrt{2-\sqrt{2}}$ | $-\sqrt{\dfrac{2+\sqrt{2}}{2-\sqrt{2}}}$ |

## PROBLEMS 9.10

**1** $x = \pi/3 + 2n\pi, \frac{5}{3}\pi + 2n\pi, (2n+1)\pi/2$
**3** $x = \pi/2 + 2n\pi, \frac{7}{6}\pi + 2n\pi, \frac{11}{6}\pi + 2n\pi$
**5** $x = \pi/4 + 2n\pi, \frac{3}{4}\pi + 2n\pi, \frac{5}{4}\pi + 2n\pi, \frac{7}{4}\pi + 2n\pi$
**7** Write: $1/(\cos x) + 1 = 2\cos x$, that is, $2\cos^2 x - \cos x - 1 = 0$
$x = 2n\pi, \frac{2}{3}\pi + 2n\pi, \frac{4}{3}\pi + 2n\pi$
**9** Write: $\csc x \tan x = \tan x$; that is, $\tan x(\csc x - 1) = 0$. But $\tan x = 0$
and $\csc x - 1 = 0$ are ruled out. Why? No solution.
**11** $x = 0.209 + (2n+1)\pi, 2n\pi - 0.209$

## PROBLEMS 9.11

**1 a** $\pi/4$  **b** $5\pi/6$  **c** $0$  **d** $-\pi/4$
**3 a** $\pi/6 + \pi/3 = -(-\pi/2)$  **b** pythagorean triangle
**5** Hint: $\sin(2\,\mathrm{Tan}^{-1}\frac{1}{3} + \mathrm{Tan}^{-1}\frac{1}{7}) = \sin(\pi/4)$; that is,
$\sin(2\,\mathrm{Tan}^{-1}\frac{1}{3})\cos(\mathrm{Tan}^{-1}\frac{1}{7}) + \cos(2\,\mathrm{Tan}^{-1}\frac{1}{3})\sin(\mathrm{Tan}^{-1}\frac{1}{7}) = \frac{1}{2}\sqrt{2}$, etc.

## PROBLEMS 9.12

**1** One-fourth of circumference
**3** Circumference
**5 a** $1719°$  **b** $2578°$  **c** $172°$  **d** $57°$  **e** $29°$  **f** $-19°$
**g** $-32°$  **h** $15°$
**7 a** $0.7547$  **b** $0.7986$  **c** $-0.7986$  **d** $-0.8090$
**e** $0.2756$  **f** $0.8480$
**9 a** $-57.2900$  **b** $-0.1763$  **c** $0.8391$  **d** $-0.5774$
**e** $0.9657$  **f** $2.7475$
**11 a** $39°$  **b** $150°$  **c** $240°$  **d** $348°$
**13 a** $81°$  **b** $173°$  **c** $236°$  **d** $342°$

## PROBLEMS 9.13

**1** 99.97 feet
**3 a** 29.4 inches **b** 68.8 inches
**5** $8 + \sqrt{19}$ feet

## PROBLEMS 10.2

**1** Any eight selected from $(3, 3\pi/4 + 2n\pi)$ and $(-3, 3\pi/4 + (n+1)\pi)$, where $n = 0, \pm1, \pm2, \ldots$

**3 a** $(-1, 2+\pi), (-1, 2-\pi), (1, 2-2\pi), (1, 2+2\pi)$ **b** $(\frac{3}{2}, -\frac{1}{3}\pi)$, $(\frac{3}{2}, \frac{5}{3}\pi), (-\frac{3}{2}, \frac{8}{3}\pi), (-\frac{3}{2}, -\frac{4}{3}\pi)$ **c** $(-3, 0), (-3, 2\pi), (3, \pi), (3, -3\pi)$
**d** $(0, \theta)$ $\theta$ is any angle

**5** 2: $(r, \theta)$ and $(-r, \theta + \pi)$, if $\theta \leq \pi$; or $(r, \theta)$ and $(-r, \theta - \pi)$, if $\theta \geq \pi$

**7** See Fig. A.71. **9** See Fig. A.72.

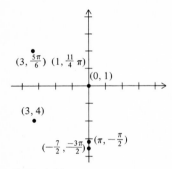

Figure A.71

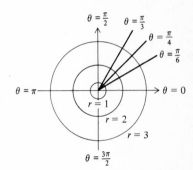

Figure A.72

## PROBLEMS 10.3

**1** See Fig. A.73. **3** See Fig. A.74. **5** See Fig. A.75.

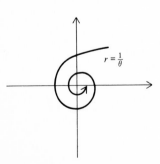

Figure A.73

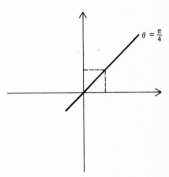

Figure A.74

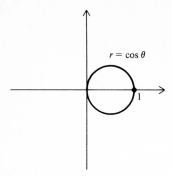

$r = \cos \theta$

1

**Figure A.75**

**7** See Fig. A.76.    **9** See Fig. A.77.    **11** See Fig. A.78.

**13** See Fig. A.79.    **15** See Fig. A.80.    **17** See Fig. A.81.

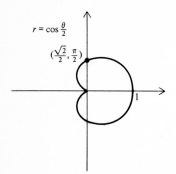

$r = \cos \frac{\theta}{2}$

$(\frac{\sqrt{2}}{2}, \frac{\pi}{2})$

1

**Figure A.76**

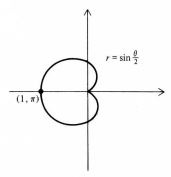

$r = \sin \frac{\theta}{2}$

$(1, \pi)$

**Figure A.77**

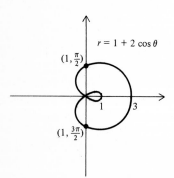

$r = 1 + 2 \cos \theta$

$(1, \frac{\pi}{2})$

1    3

$(1, \frac{3\pi}{2})$

**Figure A.78**

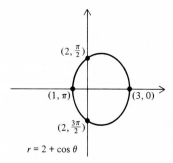

$(2, \frac{\pi}{2})$

$(1, \pi)$    $(3, 0)$

$(2, \frac{3\pi}{2})$

$r = 2 + \cos \theta$

**Figure A.79**

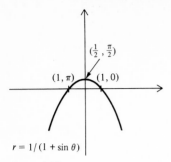

$r = 1/(1 + \sin \theta)$

**Figure A.80**

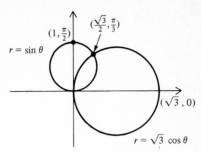

**Figure A.81**

## PROBLEMS 10.4

**1** **a** $(0, 2)$   **b** $(\sqrt{\frac{3}{2}}, -\sqrt{\frac{3}{2}})$   **c** $(-7\sqrt{3}/2, -\frac{7}{2})$   **d** $(-\pi, 0)$

  **e** $(0, 0)$   **f** $(\pi, 0)$

**3** **a** $(\sqrt{2}, \pi/4), (-\sqrt{2}, 5\pi/4)$   **b** $(\sqrt{\frac{5}{2}}, 0.466), (-\sqrt{\frac{5}{2}}, 0.466 + \pi)$

  **c** $(17, \pi), (-17, 0)$   **d** $(5\sqrt{5}, 1.108), (-5\sqrt{5}, 1.108 + \pi)$

  **e** $(5\sqrt{5}, \pi - 1.108), (-5\sqrt{5}, -1.108)$   **f** $(0, \theta)$   $\theta$ is any angle

**5** See Fig. A.82.   **7** See Fig. A.83.

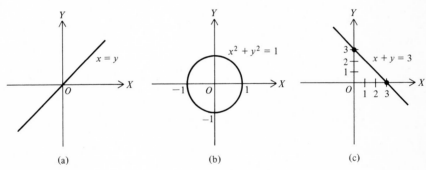

**Figure A.82**

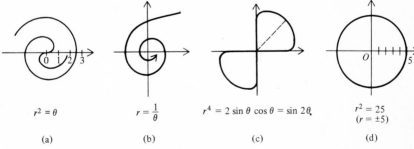

**Figure A.83**

## PROBLEMS 10.5

**1** $(-\frac{1}{4}, \frac{15}{16})$, $(0, 1)$, $(\frac{1}{2}, \frac{3}{4})$, $(\frac{5}{6}, \frac{-88}{9})$, $(1, 0)$     **3** See Fig. A.84.

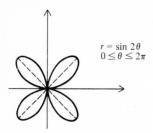

$r = \sin 2\theta$
$0 \le \theta \le 2\pi$

**Figure A.84**

## PROBLEMS 11.1

**1** See Fig. A.85.
**3.** $P(-3, \frac{3}{2}, 2)$
  $Q(1, -1, -2)$

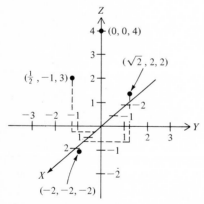

**Figure A.85**

## PROBLEMS 11.2

**1 a** $\sqrt{3}$   **b** $3$   **c** $\sqrt{\frac{4}{9} + \frac{13}{16}}$   **d** $\sqrt{16.0196}$
**3** See Fig. A.86.   **5** See Fig. A.87.
**7** $(x - a)^2 + (y - a)^2 + (z - a)^2 = a^2$, where $a > 0$

## PROBLEMS 11.3

**1** See Fig. A.88.   **3** See Fig. A.89.   **5** See Fig. A.90.

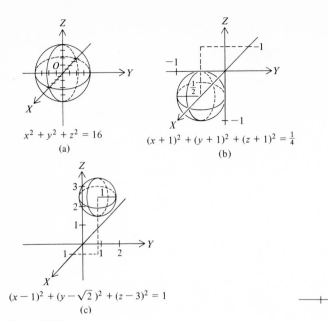

$x^2 + y^2 + z^2 = 16$

(a)

$(x + 1)^2 + (y + 1)^2 + (z + 1)^2 = \frac{1}{4}$

(b)

$(x - 1)^2 + (y - \sqrt{2})^2 + (z - 3)^2 = 1$

(c)

**Figure A.86**

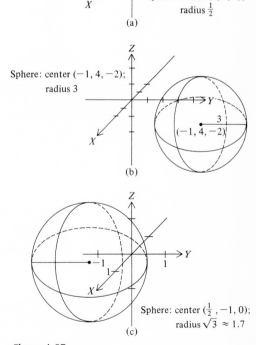

Sphere: center $(0, 0, 0)$; radius $\frac{1}{2}$

(a)

Sphere: center $(-1, 4, -2)$; radius 3

$(-1, 4, -2)$

(b)

Sphere: center $(\frac{1}{2}, -1, 0)$; radius $\sqrt{3} \approx 1.7$

(c)

**Figure A.87**

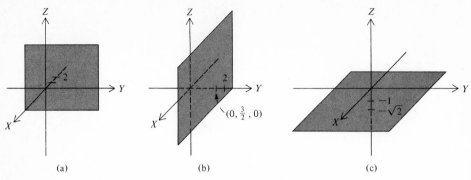

**Figure A.88**

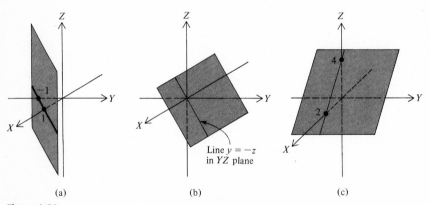

**Figure A.89**

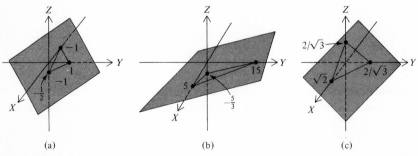

**Figure A.90**

**7**   See Fig. A.91.        **9**   See Fig. A.92.        **11**   See Fig. A.93.

## PROBLEMS 11.4

**1**   See Fig. A.94.        **3**   See Fig. A.95.        **5**   See Fig. A.96.

## PROBLEMS 11.5

**1**   See Fig. A.97.        **3**   See Fig. A.98.

**5**   See Fig. A.99.        **7**   See Fig. A.100.

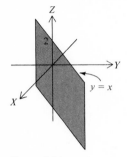

Figure A.91

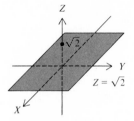

Figure A.92

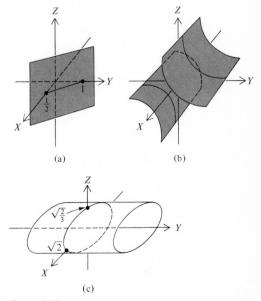

Figure A.94

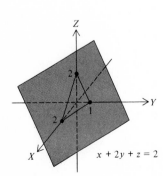

Figure A.93

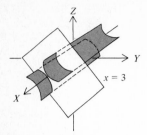

**Figure A.95**

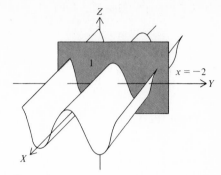

**Figure A.96**

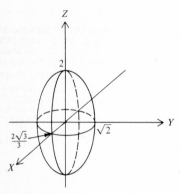

**Figure A.97**

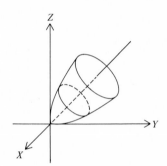

**Figure A.98**

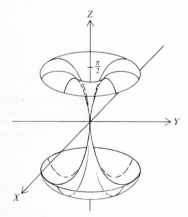

**Figure A.99**

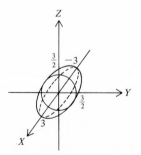

**Figure A.100**

## APPENDIX A

1    $x^3 + 3x^2 + 11x + 30 + 94/(x - 3)$

3    $x^3 + 2x^2 + 8x + 35 + 139/(x - 4)$

5    $P_1(3) = 94$      $P_2(-\frac{1}{2}) = \frac{3}{16}$      $P_3(4) = 139$      $P_4(-\frac{3}{2}) = \frac{-467}{32}$

7    $P(-0.2) = 10.831$

# INDEX

# INDEX